Bayesian Analysis of
Gene Expression Data

Bayesian Analysis of Gene Expression Data

Bani K. Mallick,

Texas A&M University, USA

David Lee Gold,

University at Buffalo, The State University of New York, USA

and

Veerabhadran Baladandayuthapani,

University of Texas MD Anderson Cancer Center, USA

⊛**WILEY**

A John Wiley and Sons, Ltd., Publication

This edition first published 2009
© 2009, John Wiley & Sons, Ltd

Registered office
John Wiley & Sons Ltd, The Atrium, Southern Gate, Chichester, West Sussex, PO19 8SQ, United Kingdom

For details of our global editorial offices, for customer services and for information about how to apply for permission to reuse the copyright material in this book please see our website at www.wiley.com.

Library of Congress Cataloguing-in-Publication Data:

Bayesian analysis of gene expression data / edited by Bani Mallick, David Gold, and Veera Baladandayuthapani.
    p. cm.
    Includes bibliographical references and index.
    ISBN 978-0-470-51766-6 (cloth)
    1. Gene expression–Statistical methods. 2. Bayesian statistical decision theory. I. Mallick, Bani K., 1965- II. Gold, David, 1970- III. Baladandayuthapani, Veerabhadran, 1976-
    QH450.B38 2009
    572.8′6501519542–dc22

2009022671

A catalogue record for this book is available from the British Library.

ISBN 978-0-470-51766-6 (HB)

Typeset in 10/12pt Times by Laserwords Private Limited, Chennai, India.
Printed and bound Great Britain by TJ International Ltd, Padstow, Cornwall.

To parents, Koushambi Nath and Bharati Mallick and his wife, Mou

Bani K. Mallick

To Marlene S. Gold

David Lee Gold

To Upali, Aarith, Aayush and my parents

Veerabhadran Baladandayuthapani

Contents

Table of Notation

$\theta \in \Theta$	unobserved scalar parameter
$\boldsymbol{\theta} = (\theta_1, \ldots, \theta_p)$	unobserved parameter vector
$\beta \in \mathcal{B}$	unobserved regression parameter
$\boldsymbol{\beta} = (\beta_1, \ldots, \beta_p)$	unobserved regression parameter vector
p	dimensionality of unobserved parameter
$\hat{\theta}, \tilde{\theta}$	point estimators of θ, as described in text
ϵ	unobserved residual
$\boldsymbol{\epsilon} = (\epsilon_1, \ldots, \epsilon_n)$	unobserved residual vector
$(\theta_1, \theta_2, \ldots, \theta_p)$	denotes elements arranged in a vector, or vectors in columns
$m \in \mathcal{M}$	index of assumed model
$g = 1, \ldots, G$	index of gene g
$i = 1, \ldots, n$	index of subjects or arrays
$k = 1, \ldots, K$	index of groups, clusters, classes, treatments
$c = 1, \ldots, C$	index of iteration in Monte Carlo chain
y	observed scalar response
$\mathbf{y} = (y_1, \ldots, y_n)$	observed response vector
$\mathbf{x}_i$	vector of observed covariates for ith subject
$\mathbf{X} = (\mathbf{x}_1, \mathbf{x}_2, \ldots, \mathbf{x}_n)$	design matrix of observed covariates
$\mathbf{X}^T$	matrix transpose
μ	mean response
$P(A|B)$	probability of event A given B
$P(y|\theta)$	likelihood of observed response y given unobserved parameter θ
$P(\theta)$	prior density of θ
π	unobserved mixture weight, $\pi \in (0, 1)$
$\boldsymbol{\pi} = (\pi_1, \ldots, \pi_K)$	vector of unobserved mixture weights, $\sum_{k=1}^{K} \pi_k = 1$
σ^2	unobserved dispersion parameter
Σ	unobserved dispersion matrix
$(\bar{x}, \hat{\sigma}_x^2)$	sample mean and sample variance of observed vector $\mathbf{x}$
t_g	t-statistic for gene g
τ	unobserved precision parameter $\tau = \sigma^{-2}$
$N(\mu, \sigma^2)$	Gaussian density, with mean μ and dispersion σ^2

Gamma(a, b)	gamma density, with shape a and rate b
IG(a, b)	inverse gamma density, with shape a and scale b
NIG(μ, σ^2, a, b)	normal inverse gamma density
U(a, b)	uniform density, with lower and upper bounds (a, b)
MixNorm(μ, σ^2, π)	mixture Gaussian density, with mixture weights $\pi_1, \ldots, \pi_K$
Dir(α)	Dirichlet density with parameters α
$(\theta_1, \theta_2, \ldots, \theta_p)^T$	denotes p elements arranged in a vector, transposed
$\mathcal{S}(\mu, \Sigma, df)$	multivariate t density, location μ, dispersion Σ, and degrees of freedom df
$\mathbf{r}, \mathbf{g}$	channel index, red and green
$\mathcal{G}$	collection of genes, $g_1, g_2, \ldots$
$\sum_{j=1}^{J} a_j$	$a_1 + a_2 + \ldots$
Beta(a, b)	beta density with parameters a,b
Dir(α)	Dirichlet density, with parameters $\alpha = \{\alpha\}_{k=1}^{K}$
DP(α, Γ)	Dirichlet Process density, with parameters α, Γ
Bernoulli(π)	Bernoulli density, with success probability π
Wishart(α, R^{-1})	Wishart density, with scale α and $p \times p$ matrix R
InverseWishart(α, R)	inverse Wishart density, with scale α and $p \times p$ matrix R
Multinomial(K, π)	multinomial density, with parameters K and π
$\mathcal{K}$	kernel function
p	p-value

1

Bioinformatics and Gene Expression Experiments

1.1 Introduction

The field of genomics has witnessed rapid growth since 1953, when Crick and Watson were credited with the discovery of the double helix structure of DNA in the cell nucleus, a discovery that opened the floodgates for great advances in research. This new understanding of the molecular basis for the code of life has enabled refinements to earlier concepts of genetics and evolution, propelling scientific discovery at a rate and scale unseen in recorded history. The last 100 years have been a golden age for science as a whole. While much remains to be learned, concurrent efforts in functional genomics are gradually piecing together an explanation of how genomic diversity is linked with the diverse characteristics of organisms. New discoveries are being made every day. The potential, for example, to treat patients with personalized medicine, or improve agricultural yields, has never been greater. These technologies are changing the way we live, and how we use them can ultimately determine how we will survive as a species in a changing environment. We must adapt and learn in a climate of global bio-uncertainty. Bioinformatics is at the crossroads of these efforts, borrowing ideas from computer science, mathematics and statistics, engineering, biology and genetics, to translate vast amounts of genomic data into quantifiable information that can change the way we process decisions and adapt to competing forces in our diverse and changing environment.

The early work in bioinformatics largely focused on string processing (sequencing) the genomes of bacteria and other microbial life forms of interest

Bayesian Analysis of Gene Expression Data   B. Mallick, D. Gold, and V. Baladandayuthapani
© 2009 John Wiley & Sons, Ltd

to researchers in public health. Larger and more ambitious efforts emerged to complete the sequencing of the human genome, with intense competition between private industry, dominated by Celera Corporation led by CEO Craig Venter, and the Human Genome Project, an international project largely funded by governments and universities. Neither of these emerged as the clear winner, as both left us with wide gaps in our understanding of the human genome, although both claimed some degree of success. Private and public efforts can be credited with greatly increasing our abilities to collect, store and process great amounts of genomic information on many species.

With completion of the human genome looming, attention has turned to investigating the links between phenotypic traits and genomic events, e.g. improving agricultural production, and to complex diseases, with the ultimate goal of mapping their molecular profiles and improving therapies by what has been come to be known as personalized medicine. Ever larger genome-wide association studies are rapidly emerging as design standards to draw inference of genetic susceptibility to disease. These efforts benefit from large investments in public infrastructure, such as the National Center for Biotechnology Information (NCBI) and the European Bioinformatics Institute.

While technological advances offer tremendous quantities of genomic information, mapping out the genomic links between phenotypes would be very expensive, even prohibitive, without the aid of high-throughput experiments, capable of screening thousands of transcripts in unison (Ramsay, 1998). These large scale experiments are made possible by the ever increasing precision of micro-technologies, enabling the measurement of more and more genes per experiment. One difficulty in printing and measuring thousands of probe sequences on a single microarray chip is that the experimental conditions for any one probe sequence are less than optimal. More costly and time-consuming technologies offer greater accuracy in measurements. Refinements in gene expression microarray experiments have benefited from experience and multidisciplinary approaches, making next-generation arrays more accurate and consistent. High-throughput gene expression microarrays are known to yield noisy and sometimes biased measurements, and statistical modeling issues are well documented in the literature (Zhang and Zhao, 2000; Alizadeh et al., 2001; Baggerly et al., 2001; Dougherty, 2001; Hess et al., 2001; Kerr and Churchill, 2001a; Ramaswamy et al., 2001; West et al., 2001; Yang et al., 2001, 2002a).

Gene expression microarray experiments have been conducted for many reasons. Basic biology interests range from functional genomics to gene discovery, e.g. infer activation/suppression of gene pathways, a topic of ongoing interest in the field of gene expression microarray analysis (Curtis et al., 2005; Jiang and Gentleman, 2007; Subramanian et al., 2007). On the clinical side, biomarker discovery is advancing rapidly, with, for example, accounts of validated cancer signatures receiving much attention (Desmedt et al., 2008; Millikan et al., 2008). Biomarkers of interest range from diagnostic, i.e. to yield improvements in early disease detection, to prognostic, i.e. choosing therapies with the aid of informative genetic signatures (Zhou et al., 2008). Advances in microarray technologies have supported and benefited from clinical trials, with routine protocols to collect and analyze candidate gene expression signatures, as part of large cohort studies (Fu and Jeffrey, 2007; Koscielny, 2008).

The future of gene expression microarray experiments is quite promising. Whole genome gene expression microarray chips make it possible to scan the entire genome of an organism for transcriptional variation. Translational research is aimed at discovering biological mechanisms, and ultimately developing new technologies from these discoveries to improve the way we live together as a species and in balance with our environment. From industry to medical engineering and bioinformatics, multidisciplinary efforts are leading to new biotechnologies that will one day replace outmoded technologies, enabling the human race to achieve a better understanding of itself and to succeed sustainably in a changing environment. Foreseen challenges include computational hurdles, dealing with the vast amount of information available, and development of tools to mine and analyze the data, to synthesize the data in a way that is useful for learning. Overcoming differences in the way diverse information is stored and analyzed is a major task, necessary for success.

1.2   About This Book

Our purpose in writing this book is to describe the existing Bayesian statistical methods in the analysis of gene expression data. Bayesian methods offer a number of advantages over more conventional statistical techniques that make them particularly appropriate to analyze these sorts of complex data. Bayesian hierarchical models allow us to borrow strength among units across a whole data set in order to improve inference. For example, gene expression experiments used by biologists to study fundamental processes of activation/suppression frequently involve genetically modified animals or specific cell lines, and such experiments are typically carried out with only a small number of biological samples. It is clear that this amount of replication makes standard estimates of gene variability unstable. By assuming exchangeability across the genes, inference is strengthened by borrowing information from comparable units.

Another strength of the Bayesian framework is the propagation of uncertainty through the model. Gene expression data is often processed through a series of steps, each time ignoring the uncertainty associated with the previous step. The end result of this process can be overconfident inference. For example, a typical analysis may begin with some kind of normalization process. The normalized data will then used for statistical analysis for gene selection, say, which ignores any uncertainty in the normalizing process. Furthermore, the selected genes may be used in a classifier, ignoring the uncertainty in the selection process. This way the analysis is usually broken down into a collection of distinct steps that fail to correctly propagate uncertainty. In a Bayesian model it is straightforward to develop integrated models which include each of these effects simultaneously, thus retaining the correct level of uncertainty on the final estimates. Detailed modeling combined with a carefully designed experiment can allow coherent inference about the unknowns (Shoemaker et al., 1999; Broet et al., 2002; Beaumont and Rannala, 2004; Wilkinson, 2007).

This book attempts to describe these hierarchical Bayesian models to analyze gene expression data. We begin in Chapter 2 with some basic biology relating to our area of interest. In particular, we begin with the principles behind microarray experiments. We discuss DNA, cDNA, RNA, and mRNA structures. We also discuss the basic microarray technology including explaining the steps involved in generating experimental microarray data.

Chapter 3 discusses inference about differential expression based on hierarchical mixture models. It starts with the Bayesian linear model which is the basic tool to develop these hierarchical models. It also describes integrated Bayesian hierarchical models, including flexible model based normalization as a part of the model itself.

In Chapter 4 we consider the problem of Bayesian multiple testing and false discovery rate analysis. We explain it within the context of decision theory and illustrate how the choice of loss functions can develop different tests.

Chapter 5 provides a review of Bayesian classification methods to classify diseases using gene expression data. We discuss both linear and nonlinear classifiers to develop flexible models to relate gene expression and disease status.

Chapter 6 concerns Bayesian hypothesis inference for gene classes.

Chapter 7 explores clustering models for gene expression data. We review different Bayesian clustering methods based on principal components analysis, mixture of Gaussians, as well as Dirichlet processes.

Chapter 8 discusses the development of Bayesian networks to uncover underlying relationships between the genes. For example, we develop a network of dependencies between the differentially expressed genes. In this chapter, we carry out further investigation to develop probabilistic models to relate genes based on microarray data.

Chapter 9 contains some advanced topics such as the analysis of time course gene expression data and survival prediction using gene expression data. Finally, for scientists who do not have Bayesian training, there are two appendices which provide basic information on Bayesian analysis and Bayesian computation.

Our book presents state-of-the-art Bayesian modeling techniques for modeling gene expression data. It is intended as a research textbook as well as valuable desktop reference. We have used much of the book to teach a three credit hour, single semester course in the Department of Statistics, Texas A&M University.

2

Gene Expression Data: Basic Biology and Experiments

2.1 Background Biology

This section contains a brief introduction to the biology of gene expression for readers with little or no biological background. For detailed information with illustrations, see Watson et al. (1987), Nguyen et al. (2002), and Tözeren and Byers (2004).

The basic unit of life in all organisms is the cell. In order to survive, several processes must be carried out by all cells, including the acquisition and assimilation of nutrients, the synthesis of new cellular material, movement and replication. Each cell possesses the entire genetic information of the parent organism. This information, stored in a specific type of nucleic acid known deoxyribonucleic acid (DNA), is passed on to daughter cells during cell division. All cells perform some common activities known as 'housekeeping processes'. Additionally, some specific activities are carried out by specialized cells. For example, muscle cells have mechanical properties, while red blood cells can carry oxygen. All these cell types have identical genes, but they differ from each other based on expressed genes.

A gene is a specific segment of a DNA molecule that contains all the coding information necessary to instruct the cell for the creation (synthesis) of functional structures called proteins, necessary for cell life processes. Hence the primary biological processes can be observed as information transfer processes, and this is crucial to mediate the characteristic features or phenotype of the cells (like cancer and normal cells).

There is another type of nucleic acid known as ribonucleic acid (RNA). RNA molecules have chemical compositions that complement DNA and are involved

Bayesian Analysis of Gene Expression Data B. Mallick, D. Gold, and V. Baladandayuthapani
© 2009 John Wiley & Sons, Ltd

in the synthesis of protein. The flow of information starts from genes encoded by DNA to a particular type of RNA known as messenger RNA (mRNA) by the transcription process and from mRNA to protein by the translation process. Hence, the kind and amount of protein present in the cell depends on the genotype of the cell. In this way genes determine the phenotype of cells and hence the organism. This simplified model to relate gene to phenotype is illustrated below:

$$\text{DNA} \rightarrow \text{mRNA} \rightarrow \text{Amino}:\text{acid} \rightarrow \text{Protein} \rightarrow \text{Cell}:\text{phenoype}$$

$$\rightarrow \text{Organism}:\text{phenotype}$$

A gene is expressed if its DNA has been transcribed to RNA, and gene expression is the level of transcription of the DNA of the gene. This is known as transcription level gene expression and microarrays measure this gene expression. We can also obtain gene expression at the protein level where the mRNA is translated to protein, and protein arrays have also been developed (Haab et al., 2001). Southern blotting and other methods can detect mRNA expression of a single or a few genes. The novelty of microarrays is that they quantify transcript levels (mRNA expression levels) on a global scale by quantifying the transcript abundance of thousands of genes simultaneously. Hence, this technique has allowed biologists to take a 'global perspective on life processes – to study the role of all genes or all proteins at once' (Lander and Weinberg, 2000).

There are three primary information transfer processes in functioning organisms: (1) replication, (2) transcription and (3) translation. We concentrate on transcription in what follows as it is directly relevant to DNA microarray technology.

2.1.1   DNA Structures and Transcription

DNA consists of four primary types of nucleotides: adenine (A), guanine (G), cytosine (C) and thymine (T). DNA exists as a double helix (Watson and Crick, 1953) where each strand is made of different combinations of the same four molecular beads, represented by A, C, G and T (Figure 2.1). Also the bonds joining the nucleotides in DNA are directional, with what are referred to as a $5'$ end and a $3'$ end. The beads in opposing strands complement each other according to base pairing combinations. The two strands of the helical chains are held together by hydrogen bonding between nucleotides at the same positions in opposite strands. The same position refers to an identical number of nucleotides from one end of the DNA. The A nucleotide on one strand always pairs with T on the other strand at the same position. The nucleotides C and G are similarly paired. Hence, the pairs (A, T) and (C, G) are known as complementary base pairs. This complementary base pair rule assures that the information stored in DNA is in duplicate. In this way we obtain complementary DNA strands. For example, if one DNA strand contains the sequence $5'$ AACTTG $3'$ at a certain location, the complementary strand will have the sequence $3'$ TTGAAC $5'$ at the same position. To generate new copies of DNA, each single strand of

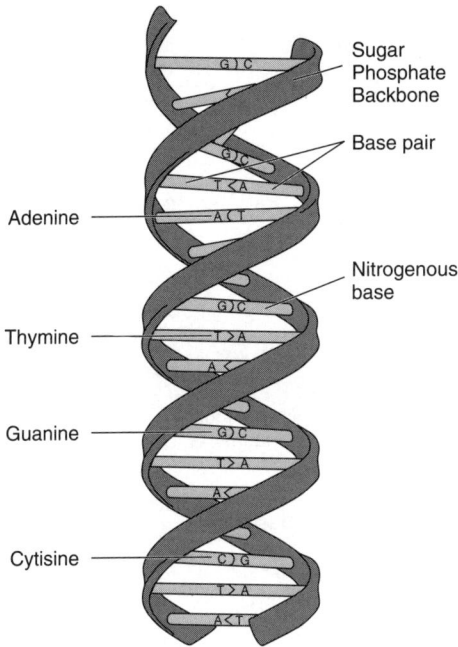

Figure 2.1   Double stranded helix structure of DNA. The types of base pairs found in DNA are restricted to those shown in the figure. (Modified from http://www.nhgri/nih.gov/DIR/VIP/Glossary.)

DNA becomes a template to produce a complementary strand. The hereditary information of an organism is distributed along both strands of a DNA molecule. The term 'gene' generally refers to those segments of one strand of DNA that are essential for the synthesis of a functional protein. It also refers to the segments of a DNA strand that encodes an RNA molecule.

RNA is a single-stranded nucleic acid containing uracil (U) in place of thymine (Figure 2.2). This uracil forms a hydrogen bond with adenine (A). DNA transcription is the information transfer process directly relevant to DNA microarray experiments because quantification of the type and amount of this copied information is the goal of the microarray experiment. In the transcription phase, the DNA sequence that ultimately encodes a protein is copied (transcribed) into RNA and this resulting RNA molecule is known as messenger RNA as it carries the information contained in DNA. There are three stages in the transcriptional process involving RNA chain: (1) initiation, (2) elongation, (3) termination.

Promoter regions are those parts of DNA that signal the initiation of transcription. Specific DNA sequences in the promoter region generate an enzyme called RNA polymerase II at the transcription initiation site. This RNA polymerase moves along the DNA and extends the RNA chain by adding nucleotides with base A, G, C, or U where T, C, G, or A is found in the DNA template strand, respectively. This complimentary pair rule is illustrated by the sequence

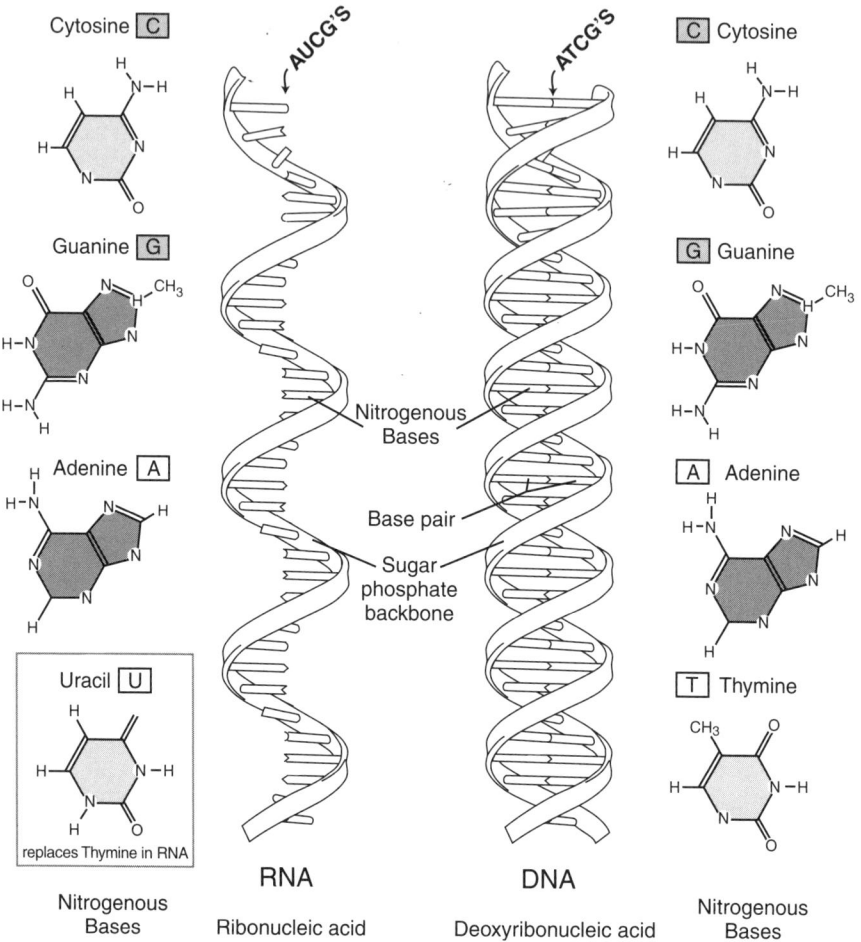

Figure 2.2   *Comparison of RNA and DNA structures. (Modified from http://www. nhgri/nih.gov/DIR/VIP/Glossary.)*

3′: TACCGAAATGAATGCGCTTA : 5′ :: DNA
5′: AUGGCUUUACUCACGCGAAU : 3′:: mRNA

RNA polymerase enzyme recognizes signals in the DNA sequences for chain termination, resulting in the release of the newly synthesized RNA from the DNA template. The resulting mRNA leaves the nucleus and associates with a ribosome, where translation occurs.

Some preprocessing usually take place before transportation of the message. For example, an enzyme called polyadenylase adds a sequence of As known as poly(A) tail to the RNA strand. The poly(A) tail plays a key role in microarray experiments. RNA splicing is another important part of mRNA formation. The

DNA segments that encode for a protein are known as exons, while noncoding segments are called introns. RNA splicing is a series of splicing reactions that remove the intron regions and fuse the remaining exon regions together. Thus, the resulting mRNA contains only the coding sequences (exons) and can be identified by the poly(A) tail.

A spliced mature mRNA molecule exists in the nucleus and is transported to a ribosome in the cytoplasm where it directs the synthesis of a molecule of protein. At the ribosome, the mRNA molecule directs the synthesis of a protein molecule via a special genetic code. It is based on triplets of contiguous nucleotides, called codons, which correspond to specific amino acids. This process of converting an mRNA molecule to a protein is called translation, and involves transfer RNA (tRNA) molecules. A unique tRNA molecule exists for each possible codon. A tRNA molecule has an anticodon at one end, which binds to its corresponding codon. At the other end, it has the corresponding amino acid. The anticodons of tRNA molecules bind to the corresponding codons of the mRNA molecule to bring the correct sequence of amino acid together. We illustrate this process in Figure 2.3.

It is clear that a correspondence exists between protein molecules and mature mRNA molecules. Mature mRNA molecules are also known as mRNA transcripts because they are synthesized by transcription of DNA. Each synthesized molecule of protein requires and consumes one transcript, hence the rate of synthesis of a protein can be estimated by quantifying the abundance of corresponding transcripts. DNA microarrays are assays for measuring the abundance of mRNA transcripts corresponding to thousands of genes in a collection of cells.

2.2   Gene Expression Microarray Experiments

Gene expression microarray technologies yield discrete measurements of thousands of transcripts at a fractional per gene basis compared to other methods, facilitating genome-wide discovery of the *relative* fold change in gene expression in, for example, diseased versus normal tissues (Alizadeh et al., 1999), discrete time course in the progression of disease (Richardson et al., 2007) or in different disease tissue types (Yeang et al., 2001). Moreover, these technologies offer the potential to discover molecular signatures that can differentiate what was once thought to be the same disease, i.e. to differentiate pathologically similar though molecularly distinct disease subtypes (Bhattacharjee et al., 2001; DeRisi et al., 1996; Weigelt et al., 2008; Eisen et al., 1998). Clinical applications include but are not limited to diagnostic and prognostic indicators (Cardoso et al., 2008; Bueno-de-Mesquita et al., 2007; Fan et al., 2006; Winter et al., 2007).

The tradeoff of measuring so many variables at once is quality. The conditions for measuring gene expression are not optimized for any one transcript, as in more expensive per gene methods. Challenges include technical artifacts, occasional poor arrays, or transcript-specific irregularities. In this section, we will examine

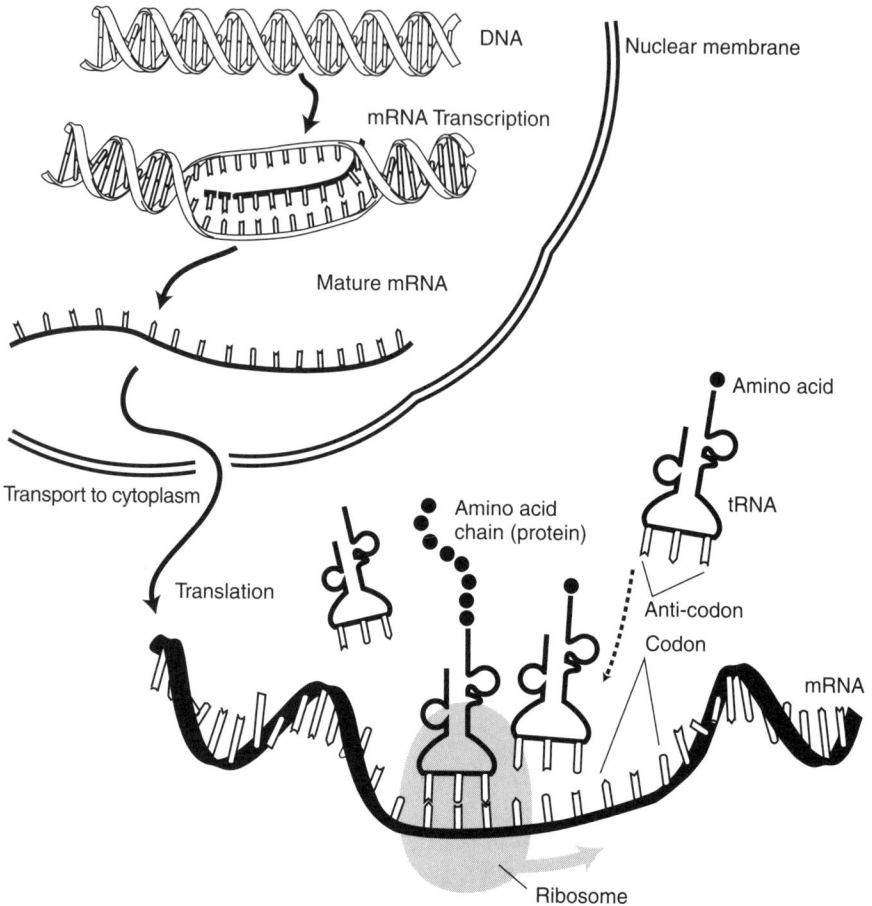

Figure 2.3   Information flow in a cell from DNA to RNA to protein. (Modified from http://www.nhgri/nih.gov/DIR/VIP/Glossary.)

some of the developments in microarray technologies, and offer insights into how the experiments are conducted.

The first account of expression profiling on a microarray was given by Schena et al. (1995). Since then, rapid investment and production of microarray experiments has been under way (Shalon et al., 1996; Schena et al., 1998; Ramsay 1998). Early experiments with gene expression microarrays largely demonstrated proof of principle (Ramaswamy et al., 2003; Shipp et al., 2002; Golub et al., 1999; Alon et al., 1999; Yeang et al., 2001). While offering promising results, early studies were plagued with technical design problems, poor array quality, and lacked reproducibility (Kerr et al., 2000; Zhang et al., 2008). Among other challenges, the genomes of many organisms were not yet fully annotated, contributing to confusion in analyzing and interpreting results.

Conventional experiments with microarrays depend on sophisticated robotics and micro-technologies. Since the advent of gene expression microarrays, powerful technologies have emerged, some commercial (e.g. Affymetrix, Agilent, Illumina) and others academic (the Stanford Brown Lab). Each technology provides its own unique results, corresponding to very different array designs and manufacturing steps. Everything from estimation of relative gene expression to high-dimensional data analysis depends on the technological platform.

2.2.1   Microarray Designs

Depending on the scientific investigator's needs, there are many microarray products to choose from, each designed to measure gene expression under different conditions, for example in *Saccharomyces cerevisiae*, *Drosophila melanogaster*, or *Homo sapiens*. Some proprietary manufacturers offer custom arrays, although in practice ready-made arrays are more commonly applied; see, for example, the Affymetrix HT Mouse Genome 430 Array or the HT Human Genome U133 Array series (http://www.affymetrix.com).

The surface of an array, glass or silicon, is made up of a grid of features, or spots, each corresponding to a gene or nucleotide sequence. Very small amounts of a DNA sequence, of the order of picomolars, of a gene are demobilized at a respective spot on the array surface. Note that the entire nucleotide sequence of each gene is not ligated to the respective spots, rather, shorter nucleotide sequences, or probes, unique to a particular chromosomal locus of a gene, are chemically attached to the chip surface. Some platforms include more than one probe per gene, and replicated probes. The genes and probes must be carefully selected in advance, requiring trial and error with clone libraries for each sequence. As some probes are better suited to microarray experimentation than others, the experimental conditions for any one probe cannot be optimized.

On an Affymetrix microarray chip, each gene is represented by a probe set, and in some cases more than one probe set per gene. Each probe set is a collection of between 14 and 20 probes. Each probe represents a short sequence, or oligonucleotide, of 25 bases called the perfect match (PM) probe. A mismatch (MM) probe, identical to the PM, with the exception of a single switch in base at the center of the nucleotide, is included for control reasons discussed below. For example, suppose that the PM probe sequence is GCACAGCTTGCAAAGGATATTGCCA. Switching the middle base from A to T, the MM probe sequence is GCACAGCTTGCATAGGATATTGCCA.

2.2.2   Work Flow

Each stage of a microarray experiment is important, relevant to the analysis, and contributes to total variation. Some steps are platform-dependent, while in general most microarray experiments include (1) sample collection, (2) preparation, (3) hybridization, and (4) fluorescent scanning and image processing. These stages of a microarray experiment are described below, with special attention to the Affymetrix platform due to its widespread application.

1. Sample Collection

Sample mRNA collection methods depend on the scientific problem, for example, from a patient *biopsy*, or from immortalized cell lines. A minimum of 5 mg of usable mRNA is generally considered sufficient and necessary for each microarray. In cases where sample is in short supply, each small quantity must be preserved. Sample testing of quality and other tissue tests for purity are unfortunately forgone in some cases, or performed on only a subsample, for want of supply. In some cases less total RNA might be considered sufficient, depending on the scientific relevance of the experiment and on the circumstances that make collection difficult or expensive.

Accepted media for tissue range from fresh or frozen tissue to paraffin embedded cores. Note that one of the inevitable challenges in working with resected tumor tissues is that the cellular makeup of the sample can vary. For example, in cancer studies, diseased tissue contains a mixture of cells, with variability in the percentage of tumor tissue present in the sample, confounding the results. Determination of the cellular content is often expensive, time-consuming and not achievable for gene expression microarray experiments. As a guide, the sample should be as homogeneous a collection of cells as possible, although in practice this cannot be guaranteed. For an in-depth study of the variability in gene expression attributable to such causes in prostate cancer, see Stuart et al. (2004).

2. Sample Preparation

The tissue sample must be prepared before it is deposited on a microarray. The mRNA is collected in a solution, fragmented with restriction enzymes, and reverse transcribed from mRNA to cloned DNA (cDNA) by reverse transcriptase catalysts. In its natural form, mRNA is relatively unstable, while cDNA is much better preserved. In some instances, amplification of the sample is performed by polymerase chain reaction. Amplification is an accepted practice for gene expression microarray experiments, although with small amounts of starting material, e.g. as little as 100 µg, experimental sensitivity is reduced (Gold et al., 2004).

The cDNA fragments are denatured with heat, so that the complementary sequences of the DNA double helix structure, see Figure 2.1, are chemically unbound, and available to bind or *hybridize* to their respective complementary oligonucleotides spotted on the array. In order to measure the relative quantity of cDNA that has hybridized at a spot, the cDNA fragments are labeled with a fluorescent dye capable of emitting a frequency pulse when activated by a laser. In *two-channel* experiments, sample is collected for two tissues of different origin, e.g. diseased and normal tissue, each labeled with a different fluorescent dye, Cy3 and Cy5 for example. Different dyes emit a different wavelength when activated, enabling measurement of the relative fluorescence at each spot. Unfortunately, different dyes are known to decay at different rates, have different dynamic ranges, and, if unaccounted for, can confound a study (Churchill 2002).

Figure 2.4   Affymetrix GeneChip® probe array. Image courtesy of Affymetrix.

Affymetrix microarray experiments are *one-channel* experiments. Short oligo sequences are labeled as cRNA rather than cDNA. The cDNAs are produced as before, but used to make target cRNAs with biotinylated nucleotides. After the sample is deposited, the chip is washed and incubated with a fluorescent dye hat binds to the biotins on the cRNAs. Dye bias is not an issue in single channel experiments.

3. *Hybridization*
A central premise underlying gene expression microarray technology is that labeled cDNA fragments will hybridize to the correct complimentary fragment on the array, a process called *complementary hybridization*. Hybridization is nature's way of attraction to and binding of complementary DNA sequences. The physics of hybridization is quite complicated. Technical spot-to-spot variability can be attributed to more than just differences in the quantity of cDNA deposited on an array or at a spot. Physical differences in the binding efficiencies of different DNA fragments are well known (Zhang et al., 2003) in target probes, as well as dye incorporation and sequence errors. Non-specific binding, whereby hybridization occurs between unintended and inexact complements, is also a confounding factor in high-throughput gene expression experiments. For this reason, Affymetrix includes a PM sequence and a MM sequence for each 25-mer oligo probe on the chip. The 25-mer mismatch is identical to the PM, except for a base pair change at the central position in the sequence. The MM signal can be used as a control to normalize the PM signal, and account for such

biases due to nonspecific binding. In practice, though, the benefit of using the mismatch probes has been called in to question, and is suspected of offering an unfavorable bias–variance tradeoff (Wang et al., 2007).

4. *Scanning and Image Processing*

The image data for a single microarray consists of pixel intensities corresponding to probes/genes as arranged on the array. When a fluorescent dye absorbs light, fluorescence is emitted, and measured by optical scanning equipment, for example the Affymetrix Scanner 3000 7G, Agilent's DNA microarray scanner with SureScan High-Resolution Technology, or Axon GenePix scanners. The scans provide a high-resolution map of the fluorescence over the array. During scanning, focused laser excitation of the dyes provides illumination over the array surface. Each pixel intensity value is stored in bits, called *color depth*. A color depth of, for example, 16 bits per pixel (as is common in microarray scanners) means that the intensity value of each pixel is an integer between 0 and $65\,535(= 2^{16} - 1)$. The number of pixels contained in a digital image is called its *resolution*. The high-resolution gray-scale TIFF image produced by the scan is used for later analysis. Note that for any given dye, there is a range of possible wavelengths. If the excitation wavelength is too close to the emission peak, the signal will be low. Too much excitation can damage the sample, resulting in a highly saturated signal, i.e. a 'bright' array. In practice, balancing the laser gain setting to optimize efficiency requires experience.

Imaging software capable of processing the array images is used to generate a detailed map of the fluorescent signal at each spot, between the spots, and over the entire array. Such software relies on sophisticated algorithms, computational signal processing tools that can identify the features on the array. *Array localization*, which is usually software-driven, involves delineating the spots corresponding to the genes in the image. Ideally, every spot should be circular in shape and all spots should have consistent diameters. This is rarely the case. The observed spots deviate from the circle in having a donut, sickle, oval or pear shape. The image analysis software rectifies these spatial problems by capturing the true shape. Other image analysis techniques use pixel distributions such as histograms to define spots. Hybrid approaches have been suggested that combine both the spatial and distributional approaches. Users may aid the software by outlining grids and providing information about spot size and the number of rows and columns spotted on the slide. Manual adjustments could be incorporated to improve upon automated spot identifications.

The pixel intensities at a spot, or feature, are a compilation of fluorescence signals from specific hybridization and all other sources. Noise in fluorescence can come from many sources, including the array surface itself, any treatments on the slide, and unintended material on the array surface. These fluorescence sources of noise must be accounted for. Only the fluorescent emission due to the biology (of scientific interest) should be included in downstream analysis. *Image segmentation* techniques are used to classify each pixel in the target area as either foreground (spot signal) or background. There are variety of proprietary commercial approaches available to this end. The spot fluorescent intensity is composed of both the biological signal we are interested in, and the background fluorescent noise, that we would like to remove. Background corresponds to the fluorescence that may contribute to the spot pixel intensities that are not due to the target molecules such as dust particles, stray molecules and the slide itself. Background areas vary across slides so most software attempts to measure local background by quantifying pixel intensities around each spot. For example, the pixel intensities between concentric circles around the target spot can be used as background. See Yang et al. (2002b) for a comparison of image analysis methods for cDNA microarrays. Systematic trends in the background fluorescence can be accounted for with spatial techniques (Wilson et al., 2003).

Having defined the spot signal and background areas, the pixels in these areas are now used to compute the spot and background intensities. Some summary statistics are used to represent the intensities for all the pixels in the respective areas. Examples include the mean (average of all pixel intensities), median, mode (peak of histogram), area (number of pixels) and total intensity (sum of pixel intensities).The best measure of intensity remains an open question, with the mean, median or some quantile of the intensity being the most common.

The imaging software used for quantification also outputs some spot quality statistics from which the reliability of the spots can be inferred. Different image analysis software includes GenePix, SPOT, ScanAlyze, UCSG Spot, Agilent's Feature Extraction software and Imagene. Many quality statistics have been proposed for measuring the quality of the signal measured at each spot. For example, the ratio of the sample mean to the sample standard deviation of the signal intensities at, i.e. pixels within or near, a spot provide a measure of the signal-to-noise for flagging weak or outlier spots.

The Affymetrix imaging software uses a griding procedure based special alignment features located in the four corners of the chip, to segment the spots. The spots, or features, on Affymetrix chips are rectangular in shape, with sides about 5–8 pixels in length. Affymetrix uses a special software to determine the locations the features on the array. The intensity value for a feature is computed by default as the 75th percentile of the pixel intensities for the whole interior of the feature, excluding the boundary pixels. If the segmentation algorithm fails to properly align the features on the chip, the signal estimates of the true features on the chip can be seriously biased. For more discussion on image processing for Affymetrix chips, see Arteaga-Salas et al. (2008) and Schadt et al. (2001) (Figs. 2.5–2.7).

2.2.3   Data Cleaning

Data cleaning protocols are universally accepted as essential for producing reasonable analyses with gene expression microarray data, although the analytical application and procedures are to a certain extent platform-specific, while general principles and methodologies can offer guidance. Nonbiological sources of variation on microarrays include technical replicate variation, dye-to-dye variation, inter-operator, array-to-array, day-to-day and inter-lab variation. Replication

RNA fragments with fluorescent tags from sample to be tested

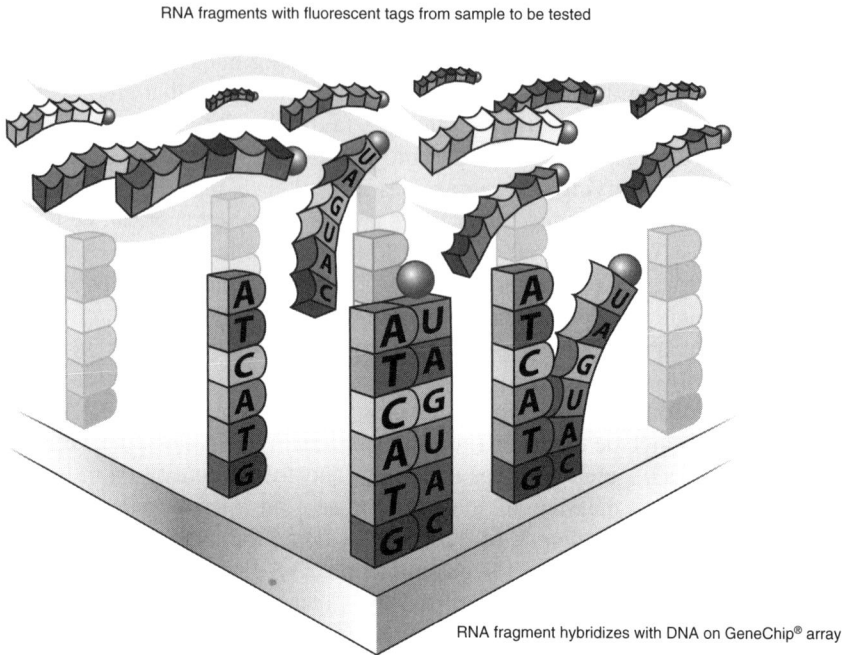

RNA fragment hybridizes with DNA on GeneChip® array

Figure 2.5   Cartoon depicting hybridization of tagged probes to Affymetrix GeneChip® microarray. Image courtesy of Affymetrix.

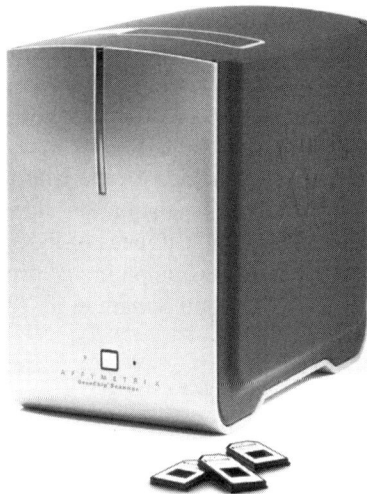

Figure 2.6   Affymetrix GeneChip® Scanner 3000. Image courtesy of Affymetrix.

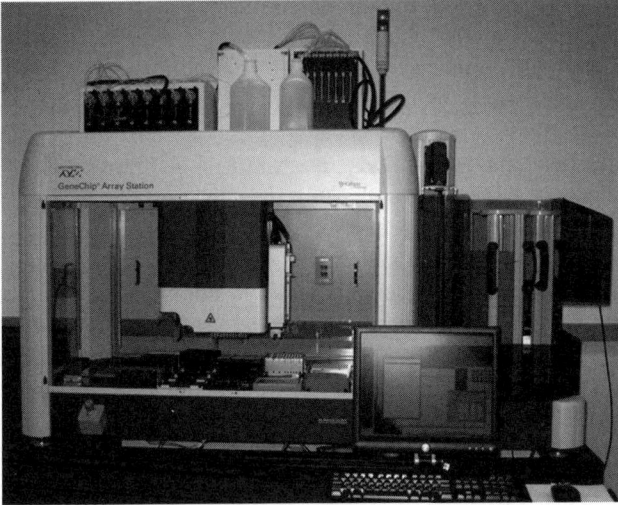

Figure 2.7   Affymetrix Array WorkStation. Image courtesy of Affymetrix.

of microarray experimental samples can provide some evidence of anomalies. Absent expensive validation techniques, there is no way of knowing if the gene expression measure for a particular gene in a given experiment is too noisy to provide information about important of biological changes.

It is almost universal practice to correct the foreground intensities by subtracting the background intensity near the probe feature. The motivation for background adjustment is the belief that a spot's measured intensity includes a contribution not specifically due to the hybridization of the target to the probe, e.g. cross-hybridization and other chemicals on the glass. An undesirable side-effect of background correction is that negative intensities may be produced for some spots and hence missing values if log-intensities are computed, resulting in loss of information associated with low channel intensities. Moreover, such a background correction tends to inflate the noise in the expression values for low expressing genes (Bolstad, 2006). A better approach is to use a smoother estimate of the background. Yang et al. (2002b) recommend a morphological background such as those produced using Spot and recent versions of GenePix. In summary, all background correction tends to increase the noise especially in low expressed genes, even if marginally, hence some users choose to ignore any sort of background correction.

Another critical step before analyzing any microarray data is array normalization. The purpose of normalization is to adjust for any bias arising from variation in the experimental steps rather than from biological differences between the RNA samples. Such systematic bias can arise, *inter alia*, from red-green bias due to differences in target labeling methods, scanning properties of two channels perhaps by the use of different scanner settings or variation over the course of the print run or nonuniformity in the hybridization. Thus it is necessary to normalize the intensities before any subsequent analysis is carried out. Normalization can be carried out within each array or between arrays. There are several methods available for array normalization such as global mean methods, iterative linear regression, curvilinear methods (e.g. lowess) and variance model methods. The simplest and most widely used within array normalization assumes that the red-green bias is constant with respect to log-scale across the array (see Ideker et al., 2000; Chen et al., 1997).

Of specific interest within a slide are systematic differences due to intensity- and location-dependent dye biases. One way to compare red and green channel measurements for each spot on an array is an MA plot. The M value on the vertical axis is the $\log_2$-ratio of the red and green channel intensities i.e. the fold change on log-scale between the samples. The A value on the horizontal axis is the average of the $\log_2$ intensities. In principle at least, it is typically expected that, free of bias, an MA plot should be centered around 0, although this is rarely that case. Trends in M versus A can be signs of channel bias. A smoother, such as a loess smoother, is usually used to give an intensity-dependent correction, such that the M values are now centered around 0. This removes any global intensity-dependent dye biases and is sometimes called the lowess normalization method. Although the lowess normalization method adjusts for location, it does not account for spatial/regional differences in variability, i.e. M values have different variability depending upon the grid they lie in. Yang et al. (2002a) proposed normalizing the median of the absolute deviation of M values.

There are various algorithms for normalization for Affymetrix data. Bolstad et al. (2003) propose a quantile normalization where the goal is to make the empirical distribution of intensities the same across arrays. The target distribution is found by averaging the quantiles for each of the arrays in the data set. Quantile normalization changes expression over many slides, i.e. changes the correlation structure of the data, and may affect subsequent analysis. A thorough comparison of quantile normalization with other methods, and its effects on variability and bias, can be found in Bolstad et al. (2003).

Another very popular method for normalization of Affymetrix array is robust multichip analysis (RMA); see Irizarry et al. (2003). It is implemented in the R package affy and is available for download from http://www.bioconductor.org/. The RMA algorithm only use the PM intensities and ignores the MM intensities from an array. The PM intnesies are background corrected on the raw intensity scale, to yield $y_{ij} = \log_2(\text{PM}_{ij} - \text{BG}_{ij})$, where BG_{ij} is an estimate of the background signal of the jth probe on the ith array . The RMA expression measure is then based on the following model, called the *multi-array probe-level model* (PLM):

$$y_{ij} = \beta_i + \alpha_j + \epsilon_{ij}$$

where i and j index the array and probe, respectively. The parameter β_j is the expression of the probe set in array i, α_j is the probe affinity effect for the jth probe in the probe set and the ϵ_{ij} are the residuals. The parameters β_i and α_j) are then estimated using robust methods such as the median Polish method (quicker) or via robust linear models.

RMA is not the only expression measure possible. A popular modification of the algorithm is known as GCRMA (Wu et al., 2004), which incorporates probe sequence information into the background correction algorithm. Li and Wong's (2001) dChip MBEI also uses a multi-array model which is multiplicative with additive errors fitted on the natural scale and a different nonlinear algorithm is used. Affymetrix also provides MAS 5.0 expression measures and more recently an algorithm called PLIER. MAS 5.0 values are typically noisy in the low-intensity range and use a simple linear scaling normalization.

Data cleaning methods, while essential, are not the focus of this book. Rather, we take the gene expression measurement from a preprocessed experiment as given, and proceed to the challenging tasks of extracting and interpreting information from the results. For further reading on data cleaning procedures, see Speed (2003), Bolstad et al. (2003), Gentleman et al. (2004), Brettschneider et al. (2008), and Fan et al. (2004). For comparisons of array technology platforms, see Thompson and Pine (2009), McCall and Irizarry (2008), and Yauk et al. (2004). Lab-to-lab variation is discussed in Irizarry et al. (2005), Fare et al. (2003), Dobbin et al. (2005), and MAQC Consortium (2006). For clinical applications, see Olson (2004), Pusztai and Hess (2004), and Ramaswamy and Golub (2002). For more infomation on Affymetrix microarrays and technical aspects of low-level data processing, see the dChip software (http://www.dchip.org; Li and Wong, 2001) and the Affymetrix white pages available at http://www.affymetrix.com/support/technical/whitepapers/sadd_whitepaper.pdf.

3

Bayesian Linear Models for Gene Expression

3.1 Introduction

The linear model is perhaps the most fundamental of all applied statistical models, encompassing a wide variety of models such as analysis of variance (ANOVA), regression, analysis of covariance (ANCOVA), and mixed effect models. A large number of such statistical models, when appropriately parameterized, can be reduced to a linear model, hence making it a rich class of models. Usually the aim of such statistical models is often to characterize the dependencies among several observed quantities. For example, how does the expression of given gene or group of genes affect a clinical phenotype, or how do gene expressions vary from the diseased group to the non-diseased group? A variety of such questions can be answered using variations and extensions of the basic linear model. In general, the ultimate goal of these analyses is to find a proper representation of the conditional distribution, $P(y|\mathbf{x})$, of an observed variable y, given a vector of observations, $\mathbf{x}$ based on random samples of y and $\mathbf{x}$. Although a complete characterization of the density $P(y|\mathbf{x})$ is generally difficult, estimation usually proceeds by restricting the space of densities to a parametric family, indexed by parameters, a route we shall follow here. We start with a brief introduction to linear models, especially in the Bayesian context, before delving into the application of linear models to gene expression data.

The basic ingredients of a linear model include a *response* or *outcome* variable y which can be continuous or discrete. The variables $\mathbf{X} = (x_1, \ldots, x_p)$ are called the *explanatory* variables (p in number) and can be continuous or discrete or

Bayesian Analysis of Gene Expression Data   B. Mallick, D. Gold, and V. Baladandayuthapani
© 2009 John Wiley & Sons, Ltd

a combination of both. For example, in typical settings discussed in this book $\mathbf{X}$ might represent the gene expression of p genes from a given microarray. The distribution of y given $\mathbf{X}$ is typically studied in the context of a set of units or experimental subjects, $i = 1, \ldots, n$, in which y_i and $x_{i1}, \ldots, x_{ip}$ are measured. We denote $\mathbf{y} = (y_1, \ldots, y_n)$ as the $n \times 1$ vector of response variables and $\mathbf{X} = (\mathbf{x}_1, \ldots, \mathbf{x}_p)$ as the $n \times p$ matrix of explanatory variables.

A variety of models with various dependence structures can be posited on the response and explanatory variables, and we consider the simplest case here: the *normal linear model*[1] where the distribution of $\mathbf{y}$ given $\mathbf{X}$ is normal with mean a linear function of $\mathbf{X}$ admitting the model

$$\mathbf{y} = \mathbf{X}\boldsymbol{\beta} + \boldsymbol{\epsilon} \qquad (3.1)$$

in which $\boldsymbol{\beta} = (\beta_1, \ldots, \beta_p)^T$ is a $p \times 1$ vector of regression coefficients and the error process, in the simplest case, is assumed to be $N(\mathbf{0}, \sigma^2 \mathbf{I}_n)$. More complicated error structures can be accommodated by this model depending on application, and we shall defer this to later. Thus, the parameters to estimate in this model are $\boldsymbol{\theta} = (\boldsymbol{\beta}, \sigma^2)$. Before we get into estimation, we refer to two implicit assumptions. First, we assume that the response variables and explanatory variables are clearly defined (with appropriate transformations if required). Second, we assume that the explanatory variables are observed without error. Further modeling mechanisms need to be in place to account for such errors, but these lie outside the scope of this book. The main practical advantage of making these assumptions is that it is much easier to specify a realistic conditional distribution on one variable $\mathbf{y}$ rather than a joint distribution on $p + 1$ variables. We note there that there is a huge literature on Bayesian analysis of linear models and it is not our intention to reproduce that here; rather, we present those aspects of estimation and inference that are pertinent to our case studies on gene expression data.

3.2   Bayesian Analysis of a Linear Model

Before discussing the Bayesian approach in detail, we briefly discuss the classical estimation of the linear model (e.g. maximum likelihood estimation), and relate the results from a Bayesian perspective. Under the Gaussian specification above, the likelihood of a simple linear model can be written as

$$l(\boldsymbol{\beta}, \sigma^2 | \mathbf{y}, \mathbf{X}) \propto (\sigma^2)^{-n/2} \exp\left\{ \frac{1}{2\sigma^2} (\mathbf{y} - \mathbf{X}\boldsymbol{\beta})^T (\mathbf{y} - \mathbf{X}\boldsymbol{\beta}) \right\}. \qquad (3.2)$$

The maximum likelihood estimator of $\boldsymbol{\beta}$ is then obtained by maximizing the above likelihood (or its logarithm, the log-likelihood), and is given by $\widehat{\boldsymbol{\beta}} = (\mathbf{X}^T \mathbf{X})^{-1} \mathbf{X}^T \mathbf{y}$. This is the ordinary least squares (OLS) estimator and can be

---

[1] We shall use the terms 'linear model' and 'linear regression' interchangeably while referring to same underlying model.

viewed as the orthogonal projection of $\mathbf{y}$ on the linear subspace spanned by the columns of $\mathbf{X}$. In order to avoid non-identifiability and uniqueness problems, we assume that $\mathbf{X}$ is of full rank, i.e., $\text{rank}(\mathbf{X}) = p + 1$. In addition, we assume $p + 1 < n$, in order for the proper estimates of $\boldsymbol{\beta}$ to exist, since $\mathbf{X}^T\mathbf{X}$ is not invertible if the condition does not hold.

Similarly, an unbiased estimator of the error variance σ^2 is given by

$$\widehat{\sigma}^2 = \frac{1}{n - p}(\mathbf{y} - \mathbf{X}\widehat{\boldsymbol{\beta}})^T(\mathbf{y} - \mathbf{X}\widehat{\boldsymbol{\beta}}) \tag{3.3}$$

and $\widehat{\sigma}^2(\mathbf{X}^T\mathbf{X})^{-1}$ approximates the covariance matrix of $\widehat{\boldsymbol{\beta}}$.

3.2.1   Analysis via Conjugate Priors

Bayesian inference and estimation of the above linear model proceeds by eliciting the prior distribution of the parameters $\boldsymbol{\theta} = (\boldsymbol{\beta}, \sigma^2)$ (Lindley and Smith, 1972). A variety of choices are at our disposal here. A class of popular priors used for building Bayesian linear models are *conjugate* priors. Noting that the likelihood function in (3.2) has Gaussian kernel linked with $\boldsymbol{\beta}$ and the inverse-gamma kernel for σ^2, a conjugate class of priors is specified as

$$\boldsymbol{\beta}|\sigma^2, \mathbf{X} \sim N\left(\boldsymbol{\mu}_\beta, \sigma^2 V_\beta\right),$$

$$\sigma^2|\mathbf{X} \sim \text{IG}\left(a, b\right), \quad a, b > 0, \tag{3.4}$$

where $\boldsymbol{\mu}_\beta$ is a p-dimensional vector and V_β is a $p \times p$ positive definite symmetric matrix. We call this the normal-inverse-gamma (NIG) prior and denote it by $\text{NIG}(\boldsymbol{\mu}_\beta, V_\beta, a, b)$, which is defined by the joint probability distribution of the vector $\boldsymbol{\beta}$ and the scalar σ^2.

The posterior distribution from the NIG prior is obtained by combining the likelihood in (3.2) and the priors specified in (3.4). This leads to the posterior distribution

$$P(\boldsymbol{\beta}|\sigma^2, \mathbf{y}, \mathbf{X}) \sim N\{\boldsymbol{\mu}_\beta^*, V_\beta^*\},$$

$$P(\sigma^2|\mathbf{y}, \mathbf{X}) \sim \text{IG}(a^*, b^*), \tag{3.5}$$

where

$$\boldsymbol{\mu}_\beta^* = (V_\beta^{-1} + \mathbf{X}^T\mathbf{X})^{-1}(V_\beta\boldsymbol{\mu}_\beta + \mathbf{X}^T)^{-1},$$

$$V_\beta^* = (V_\beta^{-1} + \mathbf{X}^T\mathbf{X})^{-1},$$

$$a^* = a + \frac{n}{2},$$

$$b^* = b + \frac{1}{2}\left[\boldsymbol{\mu}_\beta^T V_\beta^{-1}\boldsymbol{\mu}_\beta + \mathbf{y}^T\mathbf{y} - \boldsymbol{\mu}_\beta^* V_\beta^{-*1}\boldsymbol{\mu}_\beta^*\right]$$

Further, the marginal posterior distribution of β can be obtained by integrating out σ^2 from the NIG joint posterior as

$$P(\beta|X, y) = \int P(\beta, \sigma^2|X, y)d\sigma^2,$$

which results in the marginal posterior distribution being a multivariate t-density with degrees of freedom $2a^*$, mean μ_β^* and covariance $(b^*/a^*)V_\beta^*$. The probability density function of a multivariate t-density with ν degrees of freedom, mean μ and covariance V is given by

$$P(t|\nu, \mu, V) = \frac{\Gamma((\nu + p)/2)\Gamma(\nu/2)}{\pi^{p/2}|\nu V|^{1/2}}\left[1 + \frac{(t - \mu)^T V^{-1}(t - \mu))}{\nu}\right]^{(\nu+p)/2}.$$

All the information for modeling the behavior of the response as a function of the covariates is contained in these posterior distributions. Apart from the goal of understanding the behavior of the response variable y, given X, another common objective of regression analysis is prediction of the response given a new or future set of observations. Suppose we have observed a new $m \times p$ matrix of regressors $\tilde{X}$, and we wish to predict the corresponding outcome $\tilde{y}$. Observe that if β and σ^2 are known, then the probability law for the predicted outcomes would be described as $\tilde{y} \sim N(\tilde{X}\beta, \sigma^2)$ However, these parameters are not known and information about them is summarized through our posterior distributions in (3.5). Hence, all predictions for the data follow from the *posterior predictive* distribution as

$$P(\tilde{y}|y) = \int P(\tilde{y}|\beta, \sigma^2)P(\beta, \sigma^2|y)d\beta d\sigma^2,$$

which, using NIG($\mu_\beta^*, V_\beta^*, a^*, b^*$) in (3.5) and after some algebra, can be shown to follow a multivariate t-density with $2a^*$ degrees of freedom, mean $\tilde{X}\mu_\beta^*$ and variance $(b^*/a^*)(I + \tilde{X}V_\beta^*\tilde{X}^T)$. Notice that there are two sources of uncertainty in the posterior predictive distribution. First, the fundamental source of variability in the model due to σ^2, unaccounted for by $\tilde{X}\beta$, and second, the posterior uncertainty in β and σ^2 as a result of their estimation from a finite sample y. It can be shown that, as the sample size $n \to \infty$, the variance due to the posterior uncertainty disappears, but the predictive uncertainty remains.

What we have discussed so far is a general setting for conducting Bayesian inference on linear models by eliciting conjugate priors. As with any Bayesian analysis, the amount of information one wishes to impart via the prior is in some sense subjective. We discuss some other choices of priors below, varying by the amount of information imparted to the posterior analysis.

A general class of noninformative priors are Jeffrey's priors, which have a close connection to the classical estimation theory. These priors can be obtained by letting $V_\beta \to 0$ (i.e. the null matrix, essentially no prior information) and

$a \rightarrow -p/2$ and $b \rightarrow 0$. This leads to the noninformative priors $P(\boldsymbol{\beta}) \propto 1$ and $P(\sigma^2) \propto \sigma^{-2}$ or, equivalently,

$$P(\boldsymbol{\beta}, \sigma^2 | \mathbf{X}) \propto \sigma^{-2}.$$

One can easily show that this prior corresponds to Jeffrey's prior with respect to the parameters (we leave it as an exercise for the reader). Note that the two distributions above are not valid probabilities, since they do not integrate to any finite number and hence are *improper* priors.[2] However, the posterior distribution is proper. It can be shown the the posterior distributions of $\boldsymbol{\beta}$ and σ^2 are

$$P(\boldsymbol{\beta} | \sigma^2, \mathbf{y}, \mathbf{X}) \sim N \left\{ \widehat{\boldsymbol{\beta}}, \sigma^2 (\mathbf{X}^T \mathbf{X})^{-1} \right\},$$

$$P(\sigma^2 | \mathbf{y}, \mathbf{X}) \sim IG \left\{ \frac{(n-p)}{2}, \frac{(n-p)\widehat{\sigma}^2}{2} \right\},$$

where $\widehat{\boldsymbol{\beta}}$ is the standard least squares estimator defined above, IG is the inverse-gamma distribution and $\widehat{\sigma}^2$ is as defined in (3.3). There are striking similarities between Bayesian estimation with Jeffrey's prior and the classical estimation. First, note that the posterior expectation of $\boldsymbol{\beta}$ is $E[\boldsymbol{\beta}|\mathbf{y}, \mathbf{X}] = \widehat{\boldsymbol{\beta}}$, which is exactly the OLS estimate, and that the (conditional) variance is given by $\sigma^2(\mathbf{X}^T\mathbf{X})^{-1}$, which can be approximated by setting $\sigma^2 = \widehat{\sigma}^2$. Second, the distribution of $\widehat{\sigma}^2$ is characterized as $(n-p)\widehat{\sigma}^2/\sigma^2$ following a chi-square distribution.

A middle-ground solution between informative and noninformative priors can be obtained by setting $V_\beta = c(\mathbf{X}^T\mathbf{X})^{-1}$ and $P(\sigma^2) \propto \sigma^{-2}$, which is commonly referred to as Zellner's g-prior: a (conditional) Gaussian prior on $\boldsymbol{\beta}$ and improper prior on σ^2. Note that it appears the the prior is data-dependent, but this is not the case since the entire model is conditional on $\mathbf{X}$. The constant c here is interpreted as a measure of the amount of information available in the prior relative to the sample. For example, setting $1/c = 0.5$ gives the prior the same weight as 50% of the sample.

3.2.2   Bayesian Variable Selection

In an ideal setting, when building a regression model, one should include all relevant pieces of information available from the data. This includes, in the regression context, all the predictor variables ($\mathbf{X}$) that might possibly explain the variability in the response variable $\mathbf{y}$. In a gene expression data context we have potentially hundreds of thousands of genes as predictor variables, which in almost all cases is much greater than the number of samples/arrays, i.e. $p \gg n$. There are several potential drawbacks when using a plain vanilla regression model with many many predictors. First, the constraint $p < n$ is of course violated when,

---

[2]*Improper* priors are the class of of priors which do not integrate to a finite number i.e. $\int f(\theta)d\theta = \infty$ with respect to a σ-finite measure.

say, using noninformative priors. Second and more importantly, including such a large number of predictors leaves little information available to obtain precise estimators of the regression coefficients with little gain in explanatory power. As is often the case, most of the genes/probes are mostly 'housekeeping' genes and have little effect on the response, and including them in our regression model results in little or almost no gain in power while making it harder to estimate the regression parameters. Thus it is important to be able to decide which variables to include from this potentially large pool of genes in order to balance good explanatory power with good estimation performance, a problem classically referred to as *variable* selection.

The variable selection problem is usually posed as a special case of the model selection problem, where each model under consideration corresponds to a distinct subset of $\mathbf{X} = (x_1, \ldots, x_p)$. In this section we discuss variable selection approaches in the context of linear multiple regression with normal likelihood specifications. Many problems of interest can be posed as linear variable selection problems. For example, in nonparametric function estimation, the values of the unknown function are represented by $\mathbf{y}$, and linear bases such as wavelets or splines are represented by $\mathbf{X} = (X_1, \ldots, X_p)$, The problem of then finding a parsimonious approximation to the function is then the linear variable selection problem.

Variable selection is essentially a decision problem, in which all potential models have to be considered in parallel against a criterion that ranks them. Formally, with p predictor variables we have potentially 2^p models to choose from, each having q predictor variables (including the null model with $q = 0$). Formally, assuming the linear regression setup as in (3.1) with $\mathbf{y} \sim N(\mathbf{X}\boldsymbol{\beta}, \sigma^2 I)$, where $\mathbf{X} = (X_1, \ldots, X_p)$, $\boldsymbol{\beta}$ is $p \times 1$ vector of unknown regression coefficients, and σ^2 is an unknown positive scalar. The variable selection problem then proceeds to identify subsets of predictors with regression coefficients small enough to ignore them. We shall describe below different Bayesian formulations of this problem distinguished by their interpretation of how small a regression coefficient must be to ignore X_j, the jth predictor (or gene). A convenient way of representing each of the 2^p subsets is via the vector

$$\boldsymbol{\gamma} = (\gamma_1, \ldots, \gamma_p)^T,$$

where $\gamma_j = 1$ or 0 according to whether β_j is included or excluded from the model. Thus $q_\gamma = \sum_{j=1}^p \gamma_j$ indexes the dimensionality of the model implied by the vector $\boldsymbol{\gamma}$.

3.2.3   Model Selection Priors

For the specification of the model space prior, most Bayesian approaches use independence priors of the form

$$P(\boldsymbol{\gamma}) = \prod_j \omega_j^{\gamma_j} (1 - \omega_j)^{1-\gamma_j}.$$

Under this prior, each predictor X_j enters the model independently of the others, with probability $p(\gamma_j = 1) = \omega_j = 1 - P(\gamma_j = 0)$. The parameter ω_j can be viewed as the (prior) weight for variable X_j, with small ω_j downweighting predictors which are of little interest. Moreover, such priors are easy to specify and can substantially reduce the computational burden since the resulting posteriors are often obtained in closed form. See, for example, Clyde et al. (1996), George and McCulloch (1993, 1997), Raftery et al. (1997), and Smith and Kohn (1996). A more useful reduction is obtained by setting $\omega_i \equiv \omega$, yielding

$$P(\boldsymbol{\gamma}) = \omega^{q_\gamma}(1 - \omega)^{p-q_\gamma}, \tag{3.6}$$

in which case the hyperparameter ω is the a priori expected proportion of X_is in the model. Setting ω to a small number yields parsimonious models. Alternatively, one could assume a prior on ω. A popular choice is a beta prior, $\omega \sim \text{Beta}(a, b)$, which yields

$$P(\boldsymbol{\gamma}) = \frac{B(a + q_\gamma, b + p - q_\gamma)}{B(a, b)}$$

where $B(a, b)$ is the beta function. A limiting case can be obtained by setting $\omega = 1/2$, giving $P(\boldsymbol{\gamma}) = 1/2^p$, which imposes a uniform prior, and puts most of its mass near models of size $q_\gamma = p/2$. More generally, one could simply put a prior $g(q_\gamma)$ on the model dimension,

$$P(\boldsymbol{\gamma}) = \left(\frac{p}{q_\gamma}\right)^{-1} g(q_\gamma),$$

of which the above two are special cases. Note that under this prior the components of $\boldsymbol{\gamma}$ are exchangeable but not independent, except in the special case of (3.6).

3.2.4   Priors on Regression Coefficients

Conditional on γ, the regression model (3.1) can now be written as

$$P(\mathbf{y}|\boldsymbol{\beta}_\gamma, \sigma^2, \boldsymbol{\gamma}) = N(\mathbf{X}_\gamma\boldsymbol{\beta}_\gamma, \sigma^2 I) \tag{3.7}$$

where $\mathbf{X}_\gamma$ is the $n \times q_\gamma$ matrix wholes columns correspond to the γth subset of $X_1, \ldots, X_p$, $\boldsymbol{\beta}_\gamma$ is a q_γ-dimensional vector of unknown coefficients, and σ^2 is the residual error variance. Note that the same σ^2 is shared by all potential models indexed by $\boldsymbol{\gamma}$, but this is something of a mathematical trick rather than being justified by the model; the independence of σ^2 and $\boldsymbol{\gamma}$ allows convenient posterior calculations. We now concentrate on eliciting priors on the regression parameters $\boldsymbol{\beta}_\gamma$ and σ^2.

The most commonly applied prior form for this setup, especially for high-dimensional problems, is the conjugate NIG prior discussed in

Section 3.2.1. Specifically, the prior consists of a conjugate q_γ-dimensional normal prior on $\boldsymbol{\beta}_\gamma$ and an inverse-gamma prior on σ^2:

$$P(\boldsymbol{\beta}_\gamma | \sigma^2, \boldsymbol{\gamma}) = N(\boldsymbol{\mu}_\gamma, \sigma^2 V_\gamma),$$

$$P(\sigma^2 | \boldsymbol{\gamma}) \sim \text{IG}(a, b), \quad a, b > 0. \tag{3.8}$$

Note that this prior, coupled with the prior $P(\boldsymbol{\gamma})$ defined in the previous section, can be viewed as mixture prior on the elements of $\boldsymbol{\beta}$. It implicitly assigns a point mass at zero for the coefficients that are not contained in $\boldsymbol{\beta}_\gamma$, and a normal prior for the nonzero coefficients. Such a setup has been used in various variable selection settings (see Chipman et al., 2001) . Also one of the first Bayesian variable selection treatments of the setup in (3.7) was used by Mitchell and Beauchamp (1988) who proposed the spike-and-slab priors, obtained by replacing the point mass at zero prior by a normal distribution with a small variance. This parametric variable selection setup was further extended by Ishwaran and Rao (2003, 2005) for gene expression data, details of which are given in Section 3.4.

A desirable property of the prior specification in (3.8) is that the full conditionals of the regression parameters $\boldsymbol{\beta}_\gamma$ and σ^2 given $\boldsymbol{\gamma}$ are available in closed form. Furthermore, we can analytically integrate them out from the joint posterior $P(\boldsymbol{\beta}_\gamma, \sigma^2, \boldsymbol{\gamma} | \mathbf{y})$ to yield the marginal density of the data in closed form,

$$P(\mathbf{y} | \boldsymbol{\gamma}) \propto |\mathbf{X}_\gamma^T \mathbf{X}_\gamma + V_\gamma^{-1}|^{-1/2} |V_\gamma|^{-1/2} (b + S_\gamma^2)^{-(n+a)/2},$$

where

$$S_\gamma^2 = \mathbf{y}^T \mathbf{y} - \mathbf{y}^T \mathbf{X}_\gamma (\mathbf{X}_\gamma^T \mathbf{X}_\gamma + V_\gamma^{-1})^{-1} \mathbf{X}_\gamma^T \mathbf{y}.$$

As we will show in subsequent chapters, the use of closed-form expressions can substantially improve posterior evaluation and speed up Markov chain Monte Carlo (MCMC) calculations. What now remains is the elicitation of the hyperparameters, $\mu_\gamma, V_\gamma, a, b$ which are crucial to the performance of the Bayesian variable selection approach described above. For small numbers of predictors sometimes prior subjective knowledge is available to elicit these parameters (Garthwaite and Dickey, 1996). But for the cases considered in this book, involving gene expression data, the number of predictors is very large and in most cases little prior knowledge is available.

For the prior mean on the regression coefficients μ_γ, it is common to set $\mu_\gamma = 0$, which corresponds to the standard Bayesian approaches to testing point null hypotheses, where under the alternative the prior is typically centered at the point null value. For choosing the prior covariance matrix, V_γ, a common specification employed is $V_\gamma = c(\mathbf{X}_\gamma^T \mathbf{X}_\gamma)^{-1}$ or $V_\gamma = c\mathbf{I}_{q_\gamma}$, where $\mathbf{I}_{q_\gamma}$ is an identity matrix of dimension q_γ and c is a positive scalar. The former choice $V_\gamma = c(\mathbf{X}_\gamma^T \mathbf{X}_\gamma)^{-1}$ serves to replicate the covariance structure of the likelihood, and yields the common g-prior recommended by Zellner (1986). The latter choice, $V_\gamma = c\mathbf{I}_{q_\gamma}$, corresponds to the belief that a priori the components of $\boldsymbol{\beta}_\gamma$

are conditionally independent. Having fixed $\mathbf{V}_\gamma$, c should be chosen large enough so that the prior is relatively flat over the region of plausible values of $\boldsymbol{\beta}_\gamma$. With $\mathbf{V}_\gamma = c(\mathbf{X}_\gamma^T \mathbf{X}_\gamma)^{-1}$, Smith and Kohn (1996) recommend $c = 100$ and report that the performance was insensitive to values of c between 10 and 10 000. This prior was also used by Lee et al. (2003) for variable selection in a classification context for gene expression data.

The above likelihood and prior setup, (3.7) and (3.8), allows for analytical marginalization of $\boldsymbol{\beta}$ and σ^2 due to their conjugate formulations. Specifically, we can integrate out $\boldsymbol{\beta}$ and σ^2 from the joint posterior $P(\boldsymbol{\beta}, \sigma^2, \boldsymbol{\gamma}|\mathbf{y})$ to yield the marginal distribution of the data which is proportional to $P(\mathbf{y}|\boldsymbol{\gamma})$, available in closed form. Specifically, the marginal distribution $P(\boldsymbol{\gamma}|\text{rest})$ can be obtained as

$$P(\boldsymbol{\gamma}|\cdot) \propto P(\mathbf{y}|\gamma) \propto P(\boldsymbol{\gamma}|\mathbf{y})$$

With the above prior formulation and with the choice of $\mathbf{V}_\gamma = c(\mathbf{X}_\gamma^T \mathbf{X}_\gamma)^{-1}$, it can shown that

$$P(\boldsymbol{\gamma}|\text{rest}) = (1 + c)^{q_\gamma/2} \{b + \mathbf{y}^T \mathbf{y} - (1 + 1/c)^{-1} \mathbf{U}^T \mathbf{U}\}^{(n+a)/2} P(\boldsymbol{\gamma})$$

where $\mathbf{U} = \mathbf{V}^T \mathbf{X}_\gamma^T \mathbf{y}$ for upper triangular $\mathbf{V}$ such that $\mathbf{V}^T \mathbf{V} = \mathbf{X}_\gamma^T \mathbf{X}_\gamma$ which can be obtained by Cholesky's decomposition. The availability of this posterior in closed form considerably speeds up subsequent MCMC calculations.

3.2.5 Sparsity Priors

The key to inducing sparsity in the number of genes selected is how we model the prior covariance matrix, $\mathbf{V}_\beta$, or in fact the elements on its diagonal, since they correspond to the conditional variances of $\boldsymbol{\beta}$. With a prior mean set to zero, small variances essentially shrink the regression coefficients to zero, thus inducing sparsity. Denote the prior on $P(\boldsymbol{\beta}) \sim N(\mathbf{0}, \mathbf{V}_\beta)$, where $\mathbf{0}$ is a p-dimensional vector of 0s. Assume independence among the βs and hence write $\mathbf{V}_\beta = \text{diag}(\lambda_1, \dots, \lambda_p)$ as a a diagonal matrix with λ_j denoting the variance of β_j. Assigning different choices of prior distributions to the λs generates different models with different degrees of sparsity to select the number of genes used. We discuss three choice of priors below.

The simplest conjugate prior formulation is obtained by setting $\lambda_j = \text{IG}(a, b)$. This model is equivalent to the automatic relevance determination model of Li et al. (2002). The hyperparameters a and b are set such that the the variance of λ_j is large. Assuming independence among the λ_j, the joint prior distribution is given by

$$\mathbf{V}_\beta = \prod_{i=1}^{p} \text{IG}(a, b).$$

Another prior that induces sparsity is the Laplace prior on β, which, as opposed to the Gaussian prior, sets nonrelevant regression coefficients exactly to zero, i.e. selection rather than shrinkage. The Laplace prior can be written as a scale mixture of normals, and can be expressed as a zero-mean Gaussian prior with an independently exponentially distributed variance. Specifically, if $P(\beta_j|\lambda_j) = N(0, \lambda_j)$ and $P(\lambda_j|\gamma) = \exp(\gamma)$, where the exponential distribution is parameterized as $x \sim \frac{\gamma}{2}\exp(\frac{\gamma x}{2})$, we obtain

$$P(\beta_j|\gamma) = \int_0^\infty P(\beta_j|\lambda_j)P(\lambda_j|\gamma)d\lambda_j.$$

Simple integral calculation yields

$$P(\beta_j|\gamma) = \text{Laplace}\left(0, \frac{1}{\sqrt{\gamma}}\right).$$

This is essentially a Bayesian version of the lasso method of Tibshirani (1996), with an added flexibility due the choices of multiple variances (λs) rather than global penalty parameters as is done in classical lasso regression.

In the absence of any information regarding the regression coefficients, a noninformative prior is Jeffrey's prior,

$$\mathbf{V}_\beta = |\mathcal{I}(\mathbf{V}_\beta)|^{1/2} = \prod_j^p \frac{1}{\lambda_j},$$

where $\mathcal{I}(\bullet)$ is Fisher's information matrix. This prior has been shown to strongly induce sparseness and yields good performance (Bae and Mallick, 2004).

Having armed ourselves with these different kinds of modeling strategies for a Bayesian linear model, we now turn our attention to applying them to gene expression data.

3.3 Bayesian Linear Models for Differential Expression

The problem of gene detection has received a great deal of attention in high-throughput gene expression analysis research. The goal of gene detection is to obtain a candidate list of genes that are reliably differentially expressed between biological populations. For instance, which genes are differentially expressed between, say, disease and normal populations? Or, which genes are differentially expressed between disease subtype populations? These are typical of the questions investigators ask when posing the problem of gene detection for a microarray gene screening analysis. The hypothesis to test, for each gene $g = 1, \ldots, p$, is that gene g is not differentially expressed between populations (H_0) against the alternative that gene g is differentially expressed between populations (H_1).

Gene detection is not always the primary objective in a microarray study, although it is often considered to be a noteworthy goal serving larger purposes in microarray experiments. For example, the goal in a study can be to develop a diagnostic predictor of disease, i.e. supervised classification, or to identify new disease subtypes, i.e. unsupervised classification. In studies such as these, obtaining a reliable list of candidate genes can lead to a better understanding of the disease, or population of interest.

In gene-screening studies, the way in which a candidate list of genes is obtained depends on the goals and costs of the study. Some studies are more liberal, while others focus more attention on controlling error rates. There is a large body of literature devoted to multiple testing, addressing these questions, specifically aimed at gene detection (Efron and Tibshirani, 2002; Muller et al., 2004; Dudoit et al., 2004; Storey et al., 2004). We will examine these and some related topics in more detail in Chapter 4.

While there have been major advances in the microarray technologies used for genome-wide screening studies, there is still much uncertainty in gene detection. Part of the reason for this uncertainty is related to variation attributed to the technology. Another important component contributing to uncertainty is incomplete knowledge about the genomes we are interested in learning about. Note that while the above hypotheses are stated for each gene, in principle the hypotheses can be, and are very often, dependent. We address both of these issues more fully, the former in Chapter 6 and the latter in Chapter 4.

The earliest notable quantitative research in microarray gene expression analysis emphasized the importance of properly accounting for variation due to the technology, the goal being to remove any systematic error introduced by the technology. A well-accepted assumption that has gained widespread importance is that the total variation in gene expression experiments may be partitioned into variation due to the technology and variation due to biology:

Total Variation = Technological Variation + Biological Variation

+ (Technological Variation)*(Biological Variation).

This expression includes a term for the interaction between technology and biology. Poor designs, lacking adequate sample replication, proved to be fatal in many early microarray studies, as the designs did not provide for estimation of technological biases affecting the results. It is not enough to assume that technological biases are negligible if they do exist. Microarray construction and experimentation involve many steps, each of which is important to the final product. Some examples of systematic technological variation are: (1) spatial variation on a single chip, (2) variation between replicate chips, (3) variation by day of experiment, and (4) probe-to-probe variation due to the physics of DNA binding affinity (Bolstad et al., 2003; Zhang et al., 2003; Scharpf et al., 2006).

It is generally agreed that variation due to the technology should be accounted for in the data analysis, although how to best achieve this is an open question. Improperly accounting for systematic variations rooted in the technology could be detrimental to an analysis. Data cleaning steps are considered fundamental to learning from microarray experiments. Heuristic approaches have grown in popularity, that is, performance of data cleaning and data modeling in separate stages. As a result, methodologies have been proposed for either data cleaning or gene detection, but seldom both. We discuss methods proposed for modeling biological variation and technological variation jointly in Section 3.7.

The hypotheses above are stated for each gene separately. A common concern in microarray analysis is that on a gene-by-gene basis, these hypotheses can be, and very frequently are, dependent. Many of the methods proposed for gene detection begin with the prior assumption of gene-wise independence. This is an assumption made largely for mathematical convenience, as it is often the case that the problem of estimating all gene-wise interactions, given a sample of size $M < N$, is intractable. Incomplete knowledge of the genomes of many species does indeed present challenges for microarray gene detection analysis. Even if gene-wise independence cannot be justified on biological grounds, other justifications are that variation attributable to gene-gene correlation is typically overwhelmed by the measurement error due to the technology (Gold et al., 2005; Reverter et al., 2005), and that historical studies relying on the independence assumption, i.e. a univariate analysis, have produced useful results (Spurgers et al., 2006). Multivariate methods have been proposed for microarray gene detection, largely concerned with statistical rather than biological properties, such as controlling error rates or improving the power for estimating interactions. We relax the assumption of gene-wise independence in Chapters 5–8. For now we take the assumption of gene-wise independence as given, with the understanding that our conclusions will be limited to our understanding of the biology.

3.3.1 Relevant Work

We discuss some of the more prominent historical research in gene detection analysis in this section. Some of these models have a Bayesian flavor although not implemented as fully Bayesian. Note that the theoretical development in many of these models began with a limited understanding of the technology.

Newton et al. (2001) proposed a gamma-gamma model for modeling gene expression in two-channel microarray experiments. For a given probe, let R and G be the observed levels of fluorescence measured in the red and green channels respectively, on one microarray. Following Newton et al.'s notation, $T = R/G$ is the red to green ratio, or fold change. The quantity of interest is $\rho = \mu_r/\mu_g$, or the true relative fold change, where μ_r and μ_g are the respective mean levels of fluorescence in both the red and green channels. R and G are assumed to follow gamma distributions, with scale parameters θ_r, θ_g and a common shape parameter a. Newton et al. further conditioned the distribution of T on the product $S = R \times G$. The distribution is

$$P(t|s, \theta, a) \propto \frac{1}{t} \exp\left\{-\theta s^{-1/2} \left(t^{1/2} + t^{-1/2}\right)\right\} \tag{3.9}$$

where θ is the common value of θ_r and θ_g, leading to the commonly observed relationship between the mean and variance in gene expression, i.e. that the variance declines exponentially as the mean signal strengthens. The means are assigned priors $\mu_r \propto 1/\theta_r$ and $\mu_g \propto 1/\theta_g$. The posterior for ρ is derived as

$$P(\rho|R, G, a, a_0, v) \propto \rho^{(a+a_0+1)} \left\{ \frac{1}{\rho} + \frac{(G + v)}{(R + v)} \right\}^{2(a+a_0)}, \tag{3.10}$$

and the Bayes estimator of ρ is

$$\hat{\rho}_B = \frac{R + v}{G + v}, \tag{3.11}$$

a shrinkage estimator, shrinking the red-green ratio to 1. Newton et al. used empirical Bayes procedures to fit their model. Differential expression is inferred by the Bayes factor

$$\text{odds} = \frac{p_A(r, g)}{p_0(r, g)} \frac{\hat{p}}{1 - \hat{p}} \tag{3.12}$$

or posterior odds, with predictive null and alternative densities $p_0(r, g)$ and $p_A(r, g)$. Prior parameters are estimated empirically, by modal values with an EM algorithm. Extensions of Newton's method have been proposed in Lo and Gottardo (2006).

While Newton et al. offered an important advance in modeling gene expression, serious challenges remained for gene detection. Microarray data exhibits a high level of noise relative to signal, and as a consequence the results are often poorly reproduced. As a result, attention turned to robust methods. Kerr et al. (2000) focused on microarray study designs and linear models for partitioning total probe variation. This was largely in response to concerns about the technology–biology interaction. Kerr et al.'s model, for two-channel experiments, included an interaction term for dye and effect:

$$\log(y_{gijk}) = \mu + A_i + D_j + V_k + G_g + (AG)_{gi} + (VG)_{gk} + \epsilon_{gijk} \tag{3.13}$$

where y_{gijk} is the expression for the gth gene in the ith array with the jth dye in the kth experimental group. Here μ is mean gene expression, A_i is the ith array effect, D_j is the jth dye effect, V_k is the overall effect of the kth experimental group, G_g is gene g's effect, $(AG)_{gi}$ is the array i by gene g interaction term, $(VG)_{gk}$ is the experimental group by gene interaction term, and ϵ_{gijk} is noise assumed to be independently and identically distributed (i.i.d.) with zero mean. Statistical inference, on, say, the experimental group–gene interaction term, $H_0 : (VG)_{g1} - (VG)_{g2} = 0$, is performed by the bootstrap, by resampling with replacement the fitted residuals, $\hat{\epsilon}_{gijk}$, and comparing the fitted differences estimated from the data, $(\widehat{VG})_{g1} - (\widehat{VG})_{g2}$, with the bootstrap distribution $(VG)^*_{g1} - (VG)^*_{g2}$. Bootstrap approaches offer advantages with noisy array measurements, although they can be computationally inefficient, depending on the objectives of the study, requiring many repeated fits of the data.

An alternative robust method requiring less computation involves permutation testing. One such method proposed for gene detection is significance analysis of microarrays (SAM). The SAM statistic is defined as

$$d(g) = \frac{\bar{x}_{I(g)} - \bar{x}_{U(g)}}{s(g) + s_0} \tag{3.14}$$

for $\bar{x}_I(g)$ and $\bar{x}_U(g)$, the sample mean gene expression levels in states I and U. In the denominator, $s(g)$ is given by

$$s(g) = \sqrt{a \left\{ \sum_m [x_m(g) - \bar{x}_I(g)] + \sum_n [x_n(g) - \bar{x}_U(g)] \right\}} \tag{3.15}$$

for $a = (1/n_1 + 1/n_2)/(n_1 + n_2 - 2)$, and $s_0 > 0$ is a user-defined constant included in the denominator to protect against underinflation of the scale. To detect significant changes in gene expression, balanced permutations of the samples are performed, by relabeling the states I and U, and recalculating (3.14), as $d_p(g)$ for $p = 1, \ldots, P$ permutations. One rationale for permutation testing is that only a moderate number of sample replicates are required to obtain a reasonable number of permutations. In Tusher et al.'s (2001) original analysis with SAM, only four sample replicates were available, allowing 36 permutations in all. Unlike univariate methods, the authors claim that SAM is robust to the independence assumption between genes. Another approach, via a mixture model formulation and fully Bayesian method, is explored by Lewin et al. (2007).

3.4   Bayesian ANOVA for Gene Selection

This section deals with Bayesian analysis of variance models and extensions to analyze gene expression data. ANOVA models are extremely popular and powerful, while being conceptually simple and intuitive. The past few years have seen a host of ANOVA models and extensions applied to microarray data[3] especially in the Bayesian paradigm. A key feature of these models is that they lend themselves nicely to a hierarchical Bayesian modeling framework in which the variability in the gene expression data can be modeled at various levels. This framework refers to a generic model building strategy, in which data (and unobserved variables) are organized into a small number of discrete levels with logically distinct and scientifically interpretable functions and probabilistic relationships between them that capture the inherent features of the data. Appendix A deals extensively with hierarchical models, and the reader is referred to that

---

[3]We will use the terms *microarray data* and *gene expression data* interchangeably throughout this book, while referring to the same data structure.

chapter for the fundamentals of the model building processes. The two main advantages of this framework are *sharing of information across parallel units* and *propogation of uncertainity through the model*, which we will illustrate in the subsequent sections of this chapter. For basic concepts and examples of hierarchical modeling and borrowing strength across units we refer to the readers to Section A.4.

As already mentioned, there is a huge literature on the use ANOVA models and their variations/extensions for gene expression data, depending on the basic scientific question one wishes to address – normalization, differential expression, feature selection, etc. The basic models extend from the log-linear model for gene expression and its extensions to more complicated settings. An outline of all the possible models is outside the scope of this chapter, but we attempt to cover a few of the most popular basic ones while referring the reader to appropriate sources for additional models. For a brief overview of ANOVA models for microarrays (more from a frequentist perspective), see Lee (2004).

ANOVA modeling is particularly attractive for high-dimensional data such as microarrays for a number of reasons. ANOVA models can be easily reparameterized as linear regression models and can borrow on the extensive machinery already in place to analyze such models such as mixed effect models and semi- or nonparametric regression tools. ANOVA modeling also has a strong basis in normal theory and thus can be applied in a variety of settings, making it a very powerful technique for applications. These features make these models extendable to complicated settings and this, combined with the hierarchical Bayesian modeling machinery, gives rise to a large class of appealing models, depending on the basic scientific questions being addressed. We will describe some of these models below.

3.4.1   The Basic Bayesian ANOVA Model

We will start with a basic one-way ANOVA model for microarray data. Let y_g be the response (expression) for single gene g, given by

$$y_g = X^T \beta_g + \epsilon_g \qquad (3.16)$$

where X^T is an (often) fixed design matrix of covariates. For example, X^T could be a matrix of indicator variables for $j = 1, \ldots, K$ treatments (often $K = 2$ in marker studies). The errors ϵ_g account for all other sources of variation and are assumed to be normally distributed as $N(0, \sigma_g^2)$. As in classical ANOVA models, we assume $\sigma_g^2 = \sigma^2$ for all g. Of primary interest is the vector of regression coefficients, $\beta_g = (\beta_{1g}, \ldots, \beta_{Kg})$, characterizing the behavior of a particular gene, g, across all treatments. Depending on the problem at a hand, a wide class of priors, both informative and noninformative, can be adopted for β_g and σ^2 as we described in Section 3.2.

Although very simple, the above model is admittedly naive, especially for such complex data structures as microarrays. First, the genes are assumed independent of each other, which we know is biologically implausible, since genes in same regulatory pathway interact with each other. Second, the assumption of constant variance of genes is very restrictive since gene expressions vary considerably. Both of these shortcomings can be easily met by the extending the model in (3.16) to more realistic settings. This can be done in a simple and elegant manner via the hierarchical Bayesian modeling machinery, as we will show below.

3.4.2   Differential Expression via Model Selection

In this section we discuss the Bayesian ANOVA for microarrays (BAM) method of Ishwaran and Rao (2003, 2005). The authors propose an extension of the ANOVA model to detect differential expression in genes within a model selection framework. The BAM approach uses a special inferential regularization known as spike-and-slab shrinkage that provides an optimal balance between total false detections and total false non-detections. The method provides an efficient way to control the false discovery rate (FDR) while finding a larger number of differentially expressed genes. See Chapter 4 for details on multiple testing and controlling the FDR. The problem of finding differentially expressed genes is recast as determining which factors are significant in a Bayesian ANOVA model. This is done using a parametric stochastic variable selection procedure first proposed by Mitchell and Beauchamp (1988), via the hierarchical model

$$Y_i | X_i, \beta, \sigma^2 \sim N(X_i^T \beta, \sigma^2), \quad i = 1, \ldots, n,$$

$$\beta_g | \gamma_g, \tau_g^2 \sim N(0, \gamma_g \tau_g^2), \quad g = 1, \ldots, G,$$

$$\gamma_g | \lambda_g \sim (1 - \lambda_g)\delta_{\gamma^*}(\cdot) + \lambda_g \delta_1(\cdot),$$

$$\lambda_g \sim U(0, 1),$$

$$\tau_g^{-2} | a_1, a_2 \sim \text{Gamma}(a_1, a_2),$$

$$\sigma^{-2} | b_1, b_2 \sim \text{Gamma}(b_1, b_2).$$

where Y_i is the response/gene expression, X_i is the G-dimensional covariate with β as the associated regression coefficients and σ^2 the measurement error. The key feature in this model is that the prior variance $v_g^2 = \gamma_g \tau_g^2$ on a given coefficient β_g has a bimodal distribution, which is calibrated via the choice of priors on τ_g^2 and γ_g. For example, a large value of v_g^2 occurs when $\gamma_g = 1$ and τ_g^2 is large, thus inducing a large values for β_g, indicating the covariate could be potentially informative. Similarly, small values of v_g^2 occur when $\gamma_g = \gamma^*$ (fixed to a pre-specified small value), which leads to shrinkage of β_g.

Under the above model formulation, the conditional posterior mean of β is

$$E(\beta|v^2, \sigma^2, Y) = (\sigma^2\Gamma^{-1} + X^TX)^{-1}X^TY,$$

where $\Gamma = \text{diag}(v_1^2, \ldots, v_G^2)$, $\tau^2 = (\tau_1^2, \ldots, \tau_G^2)$ and $Y = (Y_1, \ldots, Y_n)$. This is the (generalized) ridge regression estimate of Y on X with weights $\sigma^2\Gamma^{-1}$. Shrinkage is induced via the small diagonal elements of Γ, which are determined by the posteriors of γ, τ^2 and λ.

This variable selection framework is then extended to microarray data via an ANOVA model and its corresponding representation as a linear regression model. The two-group setting is discussed in Ishwaran and Rao (2003) and the multigroup extension is proposed in Ishwaran and Rao (2005). For purposes of illustration, we present the the the two-group setting here. For a group $l = 1, 2$, let Y_{gil} denote the gene expression from array/individual $i = 1, \ldots, n_{gl}$ of gene $g = 1, \ldots, G$. We are then interested in identifying differentially expressed genes between two groups, say, control($l = 1$) versus treatment ($l = 2$). To this end, the ANOVA model can then be written as

$$Y_{gil} = \theta_{g0} + \mu_{g0}I\{l = 2\} + \epsilon_{gil}, \tag{3.17}$$

where the errors ϵ_{gil} are assumed i.i.d. $N(0, \sigma^2)$. θ_{g0} models the mean of the gth gene in the control group. In this model those genes that are differentially expressed correspond to $\mu_{g0} \neq 0$, i.e. turned on or off depending on the sign on μ_{g0}.

A series of transformations of the data are required before the model in (3.17) can be fitted. There are two primary transformations: centering and rescaling the data. They transformed data used for downstream analysis are

$$\widetilde{Y}_{gil} = (Y_{gil} - \bar{Y}_{g1})\sqrt{n/\widehat{\sigma}_n^2}, \tag{3.18}$$

where

$$\widehat{\sigma}_n^2 = (n - p)^{-1}\sum_{gil}(Y_{gil} - \bar{Y}_{g2}I\{l = 2\} - \bar{Y}_{g1}I\{l = 1\})^2$$

is the usual unbiased (pooled) estimator of σ_0^2, $n = \sum_{g=1}^{p} n_j$ is the total number of observations, $\bar{X}_{gl}$ is mean of group l. The effect of centering is twofold: it reduces the number of parameters, hence the effective dimension, in (3.17) from $2p$ to p and also reduces the correlation between the model parameters θ_g and μ_g. The effect of rescaling is to force the variance σ^2 to be approximately equal to n, and to rescale the posterior mean values so that they can be directly compared with a limiting normal distribution. Finally, the transformed model that is fitted to the data is

$$\widetilde{Y} = \widetilde{X}^T\widetilde{\beta}_0 + \widetilde{\epsilon},$$

where $\tilde{Y}$ is a vector of expression values obtained by concatenating the values $\tilde{Y}_{gil}$ in (3.18) in a vector, β_0 are the new vector regression coefficients containing (θ_g, μ_g) at alternate positions, and $\tilde{\epsilon}$ is the vector of measurement errors. $\tilde{X}$ is the rescaled $n \times 2p$ design matrix such that the second moments are equal to 1. It is defined in such a manner that its $2g - 1$ columns consist of 0s everywhere except for values $\sqrt{n/n_g}$ placed along the n_g rows corresponding the gene g, while column $2g$ consists of of 0s everywhere except for values $\sqrt{n/n_{g1}}$ placed along the n_{g2} rows corresponding the gene g for group 2.

As a consequence of the simple construction of the design matrix $\tilde{X}$, the conditional distribution of β can be explicitly obtained. The conditional mean for μ_g is then (approximately) equal to $(\sqrt{n_{g2}}/\widehat{\sigma_n^2})(\hat{Y}_{g2} - \hat{Y}_{g1})$. with variance approximately equal to 1. Testing whether μ_g is nonzero is then conducted by comparing its values to a $N(0, n_g/n_{g1})$ distribution called the *Zcut* procedure for differential expression.

3.5   Robust ANOVA model with Mixtures of Singular Distributions

We have seen the use of Bayesian ANOVA models to identify differentially expressed genes under different conditions using gene expression microarrays. However, there are many complex steps involved in the experimental process, from hybridization to image analysis, hence microarray data often contain outliers. These outlying data values can occur due to imperfections in the experiment or experimental equipment. Gottardo et al. (2006) developed Bayesian robust inference for differential gene expression (BRIDGE) using novel hierarchical models.

The error distribution of the ANOVA models for gene expression data may not follow a normal distribution and flatter-tailed distributions are required to capture the outliers. BRIDGE follows a robust approach using a t error distribution rather than a normal distribution. Consider a simple situation where we are comparing two independent groups (control and treatment, say, or cancer and healthy conditions) and there are J replicated observations for each gene within each group. A basic ANOVA model can be developed for such data, following the notation of the previous sections, as

$$y_{gij} = \mu_{gi} + \frac{\epsilon_{gij}}{\sqrt{w_{gij}}}, \quad \text{for } g = 1, \ldots, G; i = 1, 2; j = 1, \ldots, J,$$

where g indexes the genes, i the group (there are two groups), and j the replications, μ_{gi} is the mean expression for the gth gene for the ith group, and ϵ_{gij} is the random error. The t distribution for the error has been introduced by assuming $\epsilon_{gij}|\sigma_{gi}^2 \sim N(0, \sigma_{gi}^2)$ and $w_{gij}|v_j \sim \text{Gamma}(v_j/2, v_j/2)$ where w_{gij} and ϵ_{gij} are independent. In additional to the basic ANOVA assumptions, it has been also assume that the error variances are different for the genes and groups but the same for the replications, whereas the number of degrees of freedom for the t distribution is same for the all the genes and groups but may change among replicates. An inverse-gamma prior is assigned for the error variance parameter σ_{gi}^2.

μ_{gi} is the effect of group i on gene g, and the gth gene is not differentially expressed between the two groups if these effects are equal ($\mu_{g1} = \mu_{g2}$). Gottardo et al. (2006) developed a mixture prior for μ_{gi} to use this fact to identify significant genes. Defining $\boldsymbol{\mu}_g = (\mu_{g1}, \mu_{g2})$ and $\boldsymbol{\lambda} = (\lambda_1, \lambda_2, \lambda_{12})$, then the mixture prior is

$$\boldsymbol{\mu}_g | \boldsymbol{\lambda}, \pi \sim (1 - \pi) N(\mu_{g1} | 0, \lambda_{12}^{-1}) I[\mu_{g1} = \mu_{g2}]$$
$$+ \pi N(\mu_{g1} | 0, \lambda_1^{-1}) N(\mu_{g2} | 0, \lambda_2^{-1}) I(\mu_{g1} \neq \mu_{g2}),$$

where $N(\mu_{g1} | 0, \lambda_{12}^{-1})$ means that μ_{g1} follows a zero-mean normal distribution with variance λ_{12}^{-1}. The first component corresponds to the genes that are not differentially expressed so $\mu_{g1} = \mu_{g2}$ so for the particular gene g, and the two groups share the same variance (that way it can borrow strength). Likewise, the second component corresponds to the genes that are differentially expressed ($\mu_{g1} \neq \mu_{g2}$) so we assume independent normal priors for these two components with separate variances. Conjugate gamma priors are assigned for λs. The hierarchical model is

$$y_{gij} | \mu_{gi}, \sigma_{gj}^2, w_{gij} \sim N(\mu_{gi}, \sigma_{gj}^2 / w_{gij}),$$
$$\sigma_{gj}^2 \sim \text{IG}(a_{jo}, b_{jo}),$$
$$w_{gij} \sim \text{Gamma}(v_j/2, v_j/2),$$
$$\boldsymbol{\mu}_g | \boldsymbol{\lambda}, p \sim \text{MixNorm}(0, \boldsymbol{\lambda}, p),$$
$$\lambda \sim \text{Gamma}(\cdot)$$
$$\pi \sim \text{U}[0, 1]. \tag{3.19}$$

Gibbs sampling is required to obtain realizations from the posterior distribution. Due to assignment of conjugate priors, all the complete conditionals except for μ are in explicit form using linear model theory (see Section 3.2). To obtain the complete conditional distribution for μ, we need to use the results for the mixture distribution from the example A.3 in the Appendix A. If the prior is a mixture distribution then the posterior will again be a mixture distribution with updated components. In this situation it is easy to show using our previous results that the complete conditional distribution (conditioned on all the parameters and the vector of responses $\mathbf{y}$) for $\boldsymbol{\mu}_g$ is

$$\boldsymbol{\mu}_g | \mathbf{y}, \boldsymbol{\lambda}^*, \pi \propto (1 - \pi) N(\mu_{g1} | \theta_g, \lambda_g^{*-1}) I[\mu_{g1} = \mu_{g2}]$$
$$+ \pi N(\mu_{g1} | \theta_{g1}, \lambda_{g1}^{*-1}) N(\mu_{g2} | \theta_{g2}, \lambda_{g2}^{*-1}) I(\mu_{g1} \neq \mu_{g2}).$$

The conditional posterior precisions (inverse of variance) $\lambda_g^* = \sum_{g,i,j} w_{gij}$ $\tau_{gj} + \lambda_{12}$ and $\lambda_{gi}^* = \tau_{gi} \sum_j w_{gij} + \lambda_i$ where τ is the precision $(1/\sigma^2)$ of the error distribution. The conditional posterior means are $\theta_g = \lambda_g^{*-1} \sum_{ij} w_{gij} \tau_{gj} y_{gij}$ and $\theta_{gi} = \lambda_{gi}^{*-1} \sum_i w_{gij} \tau_{gj} y_{gij}$. To simulate μ_g from this conditional distribution, we can use a MH algorithm or direct Gibbs sampler as discussed in Appendix B.

After convergence of the MCMC chains we obtain the posterior samples of the parameters. From the posterior output we can compute the marginal posterior probability of differential expression of gene g, namely $P(\mu_{g1} \neq \mu_{g2}|\mathbf{y})$. For each gene g, This probability is computed for each gene g, and Monte Carlo samples of μ are used to estimate it. With B posterior samples, we use the relative frequency formula $\frac{1}{B} \sum I[\mu_{g1}^{(l)} \neq \mu_{g2}^{(l)}]$, where $\mu_{g1}^{(l)}$, $\mu_{g2}^{(l)}$ are the values generated at the lth iteration of the MCMC and $I(\cdot)$ is the indicator function which is 1 when $\mu_{g1}^{(l)} \neq \mu_{g2}^{(l)}$.

The method can be extended for paired samples (where the conditions are related) and the details are given in Gottardo et al. (2006).

3.6 Case Study

We demonstrate the BAM methodology on the lung cancer Affymetrix microarray data set of Wachi et al. (2005). The data consist of expression values for 22 283 genes collected from ten patients, five of whom had squamous cell carcinoma (SCC) of the lung and five were normal patients. The data set is available for download at http://www.ncbi.nlm.nih.gov/geo/query/acc.cgi?acc = GSE3268. The microarray data are normalized using the RMA method (see Section 2.2.3) in R using affy. We refer the reader to Wachi et al. (2005) for further details on the pre-processing of the data.

We used the BAMarray 2.0 academic edition software available for download at http://www.bamarray.com/ (see Ishwaran et al., 2006, for further details on BAMarray). We used the default options built into the software for the entire analysis. The BAM method assumes constant variance across the groups and an automatic CART based clustering approach to variance stabilization built within the software. The adequacy of the transformation can be visualized via a V-plot (Ishwaran and Rao, 2005) as shown in Figure 3.1. If the variances have stabilized to values near 1, then plotting the group mean difference for a gene versus the corresponding absolute value of the t-statistic should give a plot with a line having constant slope. These theoretical lines are the ones shown as dotted black lines on the V-plot and the clustering of differences along these lines represents the appropriateness of the transformation.

Figure 3.2 shows the shrinkage plot for the lung cancer data set. Plotted are the posterior variances on the vertical axis and the Zcut values on the horizontal axis. As demonstrated in Ishwaran and Rao (2005), genes that are truly differentially expressed will have posterior variances converge to 1 on the far right and left of the plot. The cutoff values are determined in a data-adaptive manner by balancing

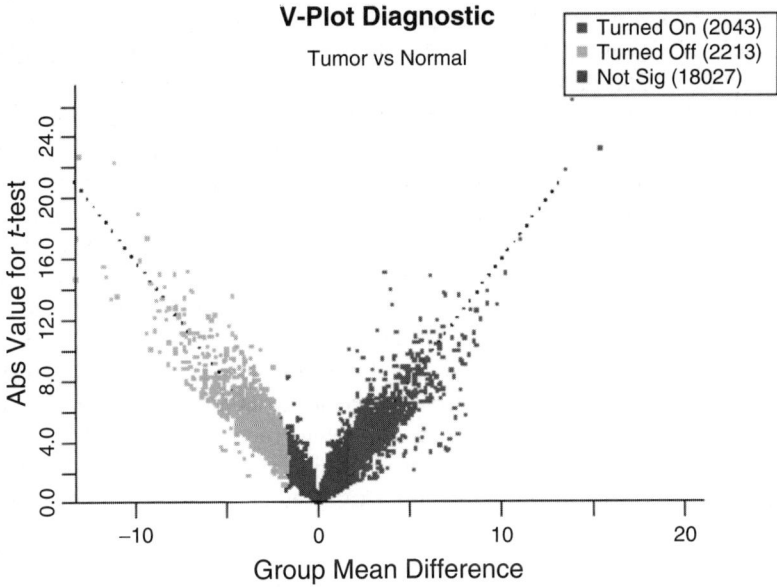

Figure 3.1   V-plot of the tumor versus normal comparison for the lung cancer data.

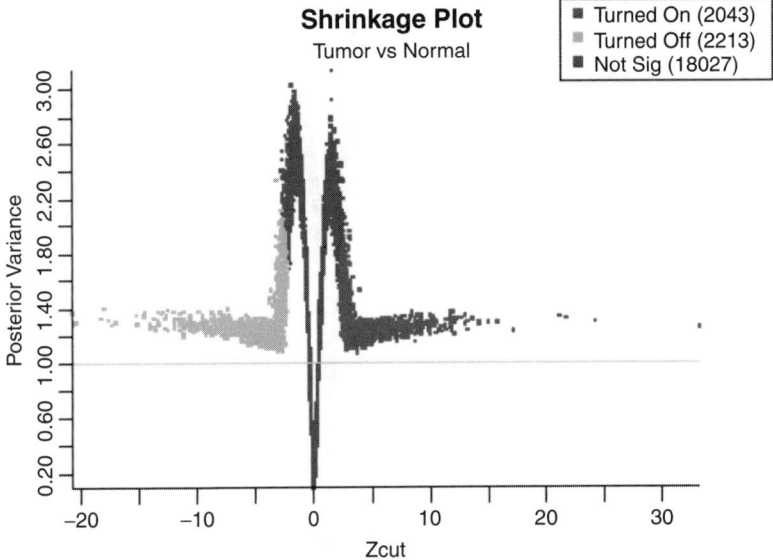

Figure 3.2   Shrinkage plot for determining differentially expressed genes for tumor groups relative to the normal group for the lung cancer data.

the total false detections against total false non-detections. It can be seen that out of the 22 383 genes: 2043 genes are turned on (up-regulated), 2213 genes are turned off (down regulated) and 18 027 genes are found not to be significant. There are thus 4256 differentially expressed genes in total.

Figure 3.3 shows the posterior probabilities from the BRIDGE method using t-distributed errors. Plotted are the posterior probabilities versus the log-posterior differences, $\log(\gamma_1 - \gamma_2)$, between the two groups. Notice how most of the log-ratios are shrunk towards zero and hence have very low posterior probability of differential expression. The BRIDGE analysis found 1982 genes differentially expressed at the 0.5 threshold for the posterior probability, which was the value the authors used in the paper. In addition, we found 1259 at 0.9, 2507 genes at 0.25, 3873 at 0.05 thresholds, as the number of differentially expressed genes. Comparing with the BAM method, we found 1741 (41%), 1191 (28%), 2101 (49%) and 2925 (69%) overlap between the number of genes that were differentially expressed. Using t-tests we found more than 10 000 genes that were differentially expressed, thus the Bayesian methods found a substantial reduction in the number of differentially expressed genes. This demonstrates the

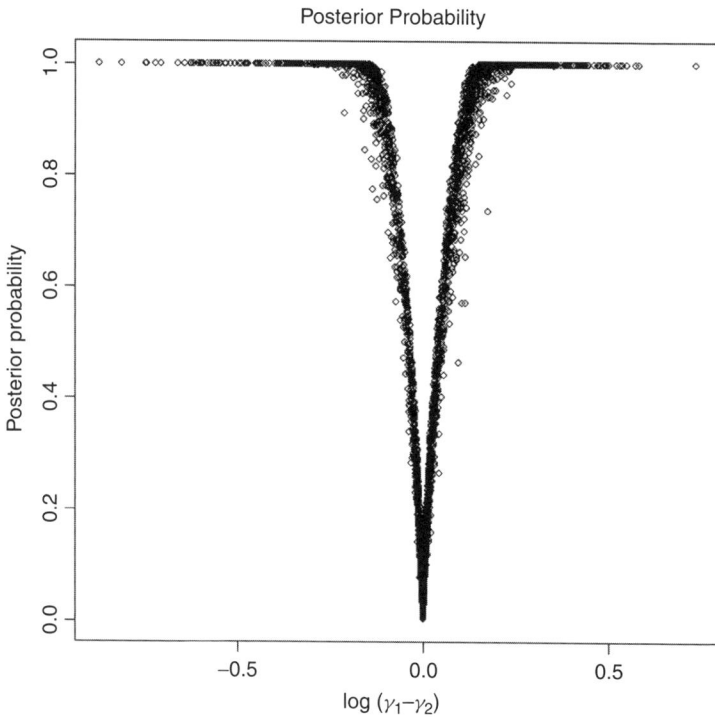

Figure 3.3   Posterior probabilities from the BRIDGE method. Plotted are the posterior probabilities on the vertical axis versus the log-posterior differences, $\log(\gamma_1 - \gamma_2)$, between the two groups.

effectiveness of the model-based Bayesian shrinkage methods, in which the posterior means are shrunk towards zero, while maintaining nominal Type I error rates. The corresponding gene-lists can be obtained from the companion website for this book. See Chapter 4 for a detailed analysis using frequentist and Bayesian FDR techniques.

3.7   Accounting for Nuisance Effects

There are many sources of signals in gene expression measurements. The signal reflection from the background of the array can contaminate the spot signal. A spot with signal below background haze is considered for practical purposes to be a measurement of noise. Background subtraction methods have been proposed, although the sensitivity to background correction is not completely understood (Scharpf et al., 2006). Low signal-to-noise ratios are commonly observed on a probe-by-probe basis in microarray analyses. If one partitions total signal into two sources,

Observed Spot Signal = True Spot Signal + Background Spot Noise,

then the goal of signal detection is to classify spots as measurements of signal from gene expression or measurements of noise, and estimate and remove the signal attributed to noise. The signal attributed to background is in this case considered to be a *nuisance parameter*, since what we want to learn about is biology. An approach similar to hard thresholding compares signal relative to an estimate of the noise in (3.7) with a user-defined threshold. A gene is called 'expressed' if the measured signal/noise is above the threshold, and unexpressed otherwise. According to this approach, genes that are consistently unexpressed are removed from the analysis, as are genes exhibiting consistently weak signal. Affymetrix produces a detection probe-wise detection p-value with their Microarray Suite software, see their White Paper (www.affymetrix.com).

Ibrahim et al. (2002) proposed a truncated model for gene expression with a component for signal detection,

$$x_g = \begin{cases} c_o, & \text{with probability } \pi_g, \\ c_o + y_g, & \text{with probability } 1 - \pi_g, \end{cases}$$

where $c_o > 0$ is the threshold below which x_g is unexpressed. The constant c_o is a user-defined minimum signal threshold. The prior probability that gene g is expressed is $\pi_g = P(x_g > c_o)$. If gene g is expressed, then x_g has a truncated distribution, where c_o is the lower bound and y_g is the continuous part, assumed to follow a lognormal distribution.

The likelihood of gene expression for gene g, y_{gij}, conditional upon individual i and tissue type k, is

$$P(y_{gik}|\mu_{gk}, \sigma_{gk}^2) = (2\pi)^{-1/2} y_{gik}^{-1} \sigma_{gk}^{-1}$$
$$\times \exp\left\{ -\frac{1}{2\sigma_{gk}^2}(\log(y_{gik}) - \mu_{gk})^2 \right\}.$$

The mean μ_{gi} is assigned a normal prior

$$\mu_{gk}|m_{k0}, \sigma_{gk}^2 \sim N(m_{k0}, \tau_o\sigma_{gk}^2/\bar{n}_k), \tag{3.20}$$

where $\bar{n}_k$ is interpreted as the total number of expressed genes across all samples in tissue type k, divided by the total number of gene expression measurements across all probes and samples. Independent inverse-gamma priors are assigned to the σ_{gk}^2. Ibrahim et al. derived the correlation between μ_{gk} and $\mu_{g'k}$ from the joint posterior

$$(\mu_{gk}, \mu_{g'k}) \sim N_2(\mu^*, \Sigma^*)$$

with $\mu^* = (m_{ko}, m_{ko})'$ and

$$\Sigma^* = \begin{pmatrix} \frac{\tau_o\sigma_{gk}^2}{\bar{n}_k} + v_{ko}^2 & v_{ko}^2 \\ v_{ko}^2 & \frac{\tau_o\sigma_{g'k}^2}{\bar{n}_k} + v_{ko}^2 \end{pmatrix},$$

implying that $\mathrm{corr}(\mu_{gk}, \mu_{g'k}|\sigma_{gk}^2, \sigma_{g'k}^2, v_{ok}) \to 1$ as $\bar{n}_k \to \infty$ or $v_{ko}^2 \to \infty$.

Their method of gene detection is based on the posterior distribution of $\xi_g = \Psi_{gk}/\Psi_{gk'}$, where

$$\Psi_{gk} = c_o p_{gk} + (1 - p_{gk})(c_o + E(y_{gik}|\mu, \sigma^2)).$$

The details of gene selection are rather involved, using what the authors call an L measure. The interested reader is referred to Appendix A.1.

In Chapter 1 we discussed the problem of array-to-array variation in detail. Methods for array normalization are largely heuristic (Bolstad), although joint modeling has been proposed. Some of the statistical methods for normalization differ depending on the technology, i.e. based on different assumptions about the underlying process that led to array-to-array disparity. A basic extension of the linear model for array normalization in single channel experiments is

$$y_{gik} = \alpha_g + \beta_{gk} + f_{ik}(\alpha_g) + \epsilon_{gik}, \tag{3.21}$$

where α_g is the mean level of gene expression for gene i, β_{gk} is the experimental effect of condition k on gene g, and ϵ_{gik} is residual noise, which has mean zero and variance σ_g^2. The term $f_{ik}(\alpha_g)$ is assumed to be a smooth function of α_g, a nuisance effect due to technological variation between the arrays. In one-channel experiments there is disagreement as to how array-to-array differences arise, although a general observation is that differences between the distributions of signal between arrays is inevitable.

A similar model may be adopted for two-channel experiments, where condition k indexes the channel and $f_{ik}(\alpha_g)$ is a channel-array dependent trend. In two-channel experiments, $f_{ik}(\alpha_g)$ is generally assumed to be the result of disparities in the measurements between the channels unrelated to biology, e.g. differences in the dynamic ranges of the fluorescent dyes, or related to differences in the manner in which the respective channels were processed, e.g. laser gain setting.

In some cases $f_{ik}(\alpha_g)$ is simply assumed to be linear. A very general assumption is that $f_{ik}(\alpha_g)$ belongs to a class of smooth functions, including both linear and nonlinear functions. The heuristic approach is to first estimate the function $f_{ik}(\alpha_g)$ on each array, $\widehat{f_{ik}(\alpha_g)}$, and then analyze the transformed response, $y_{gik} - \widehat{f_{ik}(\alpha_g)}$. A valid criticism of the heuristic approach is that real differences in the β_{gj} can distort estimation of $f_{ik}(\alpha_g)$, and therefore real experimental effects can be smoothed out in estimation and removal of array effects, $f_{ik}(\alpha_g)$. It is argued that joint estimation of β_{gk} and $f_{ik}(\alpha_g)$ does not suffer from this problem, at least not to the same degree as the heuristic approach. Iterative methods have been proposed (e.g. Huang et al., 2005; Fan et al., 2006) for two-channel experiments. In principle, these methods could be applied to one-channel array experiments.

Lewin et al. (2006) offer a fully Bayesian solution to the problem of estimating (3.21). The function $f_{ik}(\alpha_g)$ is assumed to be a linear function of second-order spline basis functions, depending on the intercept terms α_g, $g = 1, \ldots, N$:

$$f_{ik}(\alpha_g) = b_{ik}^{(0)} + b_{ik}^{(1)}(\alpha_g - a_0) + b_{ik}^{(2)}(\alpha_g - a_0)$$
$$+ \Sigma_{s=1}^{S} b_{iks}^{(2)}(\alpha_g - a_{iks})^2 I[\alpha_g \geq a_{iks}]. \qquad (3.22)$$

Here the a_{iks} are unknown knot points, while the bs are unknown polynomial spline coefficients, for an assumed fixed number of knots S. Model (3.22) as specified is not identifiable. Lewin et al. normalize within each experimental condition, such that $\sum_{k=1}^{K} f_{ik}(\alpha_g) = 0$ for all g, i, k. The authors place constrained uniform priors on the knot points and intercept terms, α_g. Vague normal priors are assigned to the coefficients. The authors considered different priors for the error variance, σ^2, a lognormal prior and a gamma prior, and compared modeling assumption (3.22) versus a piecewise constant model

$$f_{ik}(\alpha_g) = \sum_{s=1}^{S} b_{iks} \times I(a_s < \alpha_g \leq a_{a+1}), \qquad (3.23)$$

for $a_s < a_{a+1}$. Note that several very important assumptions are required in order for the model to be identifiable, and to eliminate confounding between differential gene effects, δ_g, and channel-array effects, $f_{ik}(\alpha_g)$. The channel-array effects must sum to zero, $\sum_{ik} f_{ik}(\alpha_g) = 0$, across replicates and conditions, for each gene g, and control genes (often referred to as housekeeping genes), for which

$\delta_g = 0$, must be known over a range of expression values. Care must be taken, as failure to achieve both of these requirements can lead to confounding between the differential gene and channel-array effects. With this in mind, the example below demonstrates use of Lewin et al.'s should be at their supplementary website algorithm.

Example 3.1
Two-channel microarray data were simulated, for $g = 1, \ldots, 300$ genes and $k = 1, \ldots, 5$ replicates. Channel-array effects, $f_a(\alpha_g)$ and $f_b(\alpha_g)$, were maintained the same for all replicates, between channels a and b respectively; see Figure 3.4.

Notice that for increasing mean intensity α_g, array effects are concave increasing functions, by channel, similar to the intensity-dependent saturation documented in two-channel experiments. Two simulations were performed, with and without true differential gene effects. Figure 3.5 illustrates the posterior summaries from the simulation without true differential gene effects. Differences in the channel-array effects are largely captured in the posterior means. Following the criterion, $P(\delta < -1 \cup \delta > 1 | Y) \geq 0.80$, no genes show strong evidence of differential expression.

Figure 3.6 illustrates the posterior summaries from the simulation with true differential gene effects. Gene with intensity $5 \leq \alpha \leq 10$, excluding the control

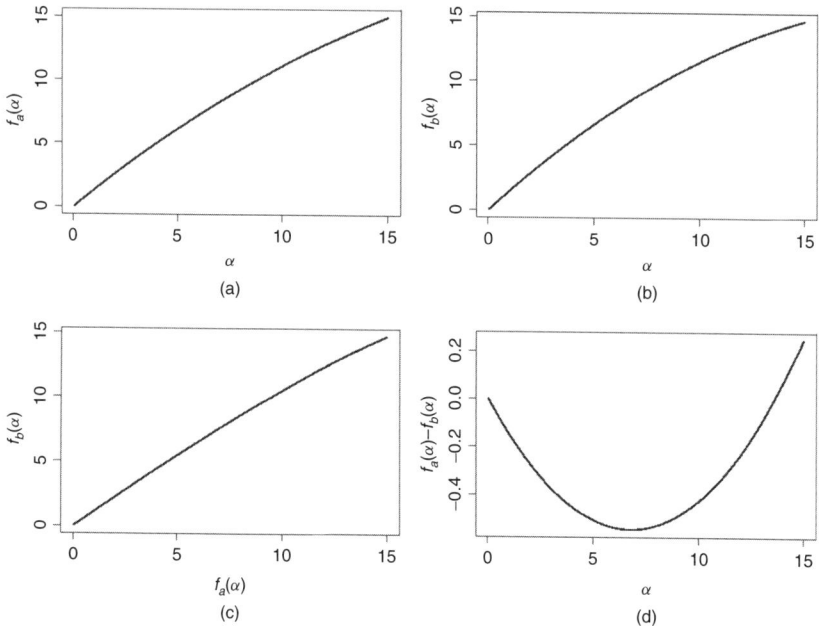

Figure 3.4   Array effects. (a) Channel a effect by mean α. (b) Channel b effect by mean α, (c) Channel a effect versus channel b effect. (d) Difference in channel effects, $f_a(\alpha) - f_b(\alpha)$ by mean α.

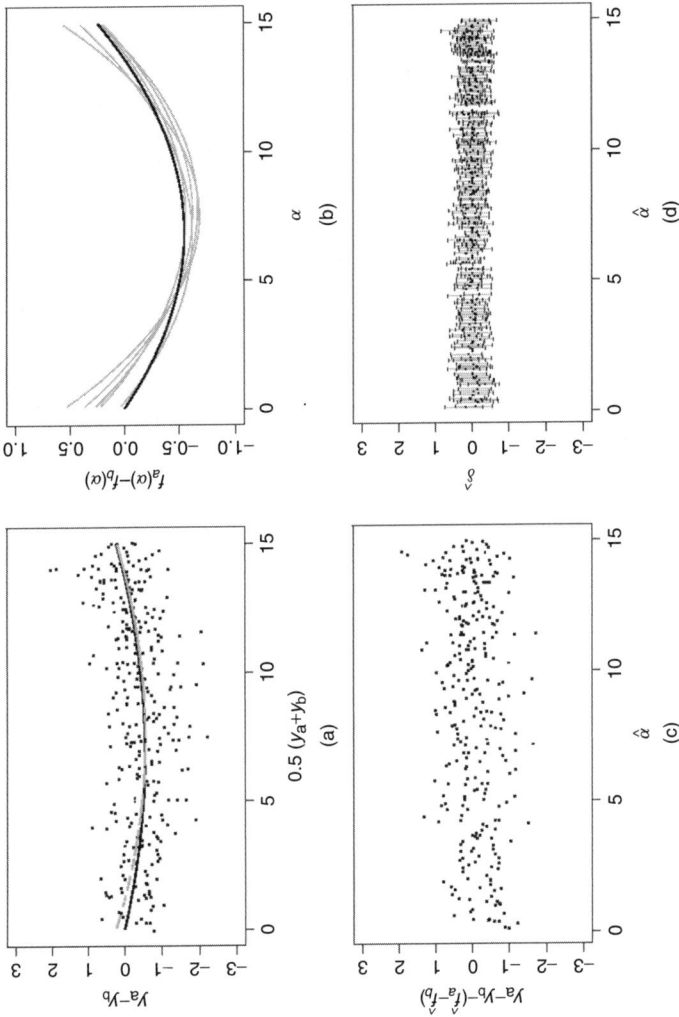

Figure 3.5 Array effects. (a) MA plot for one representative sample. Superimposed is the true difference in array effects, $f_a(\alpha) - f_b(\alpha)$, and in gray the posterior mean difference in array effect $\hat{f}_a(\hat{\alpha}) - \hat{f}_b(\hat{\alpha})$ by posterior mean gene effect $\hat{\alpha}_g$. (b) The true difference in array effects, $f_a(\alpha) - f_b(\alpha)$, is shown in black, and the posterior mean differences in array effects $\hat{f}_a(\hat{\alpha}) - \hat{f}_b(\hat{\alpha})$ are shown in gray for all five replicates. (c) MA plot for one representative sample, with the posterior mean difference in array effect, $\hat{f}_a(\alpha) - \hat{f}_b(\alpha)$, removed. (d) Posterior mean and 90% confidence interval for differential gene effects δ_g by posterior mean gene effect $\hat{\alpha}_g$.

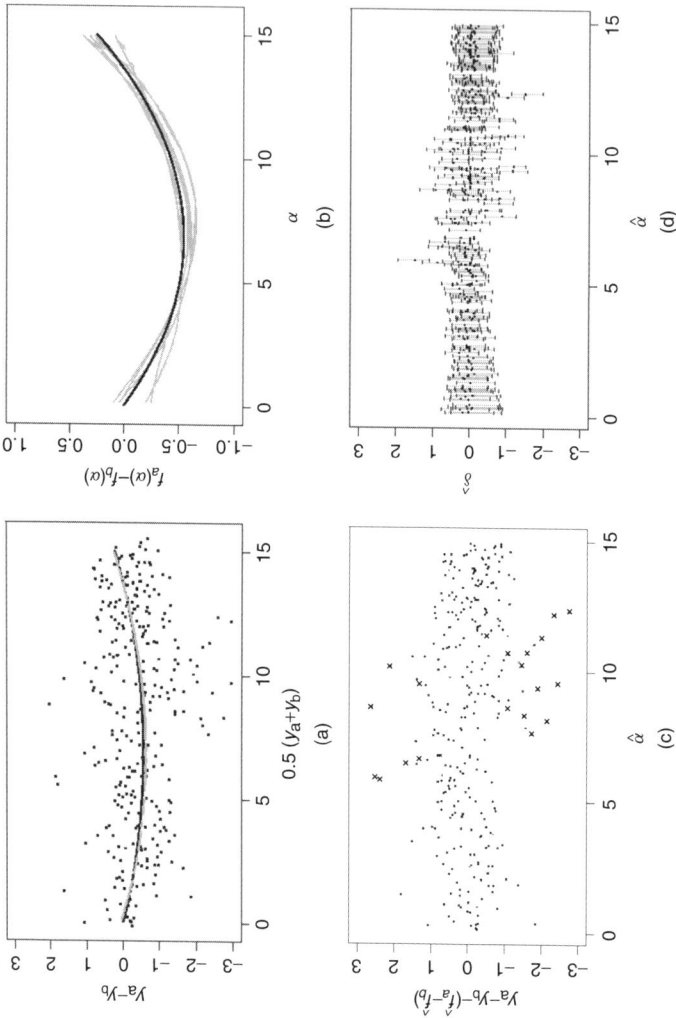

Figure 3.6   Array effects. (a) MA plot for one representative sample. Superimposed is the true difference in array effects, $f_a(\alpha) - f_b(\alpha)$, and in gray the posterior mean difference in array effect $\hat{f}_a(\hat{\alpha}) - \hat{f}_b(\hat{\alpha})$ by posterior mean gene effect $\hat{\alpha}_g$. (b) The true difference in array effects, $f_a(\alpha) - f_b(\alpha)$, is shown in black, and the posterior mean differences in array effects $\hat{f}_a(\hat{\alpha}) - \hat{f}_b(\hat{\alpha})$ are shown in gray for all five replicates. (c) MA plot for one representative sample, with the posterior mean difference in array effect $\hat{f}_a(\hat{\alpha}) - \hat{f}_b(\hat{\alpha})$ removed. Genes discovered for change are "crossed". (d) Posterior mean and 90% confidence interval for differential gene effects δ_g by posterior mean gene effect $\hat{\alpha}_g$.

genes, were given random differential gene effects following a standard normal $\delta_g \sim N(0, 1)$. Of the 89 simulated genes with true differential effects, 20 (22.47%) were discovered, following the criterion $P(\delta < -1 \cup \delta > 1|Y) \geq 0.80$. Their genes are circled in red in Figure 3.6. Note that several of the differentially expressed genes have point estimates of mean intensity above 10.

Another important nuisance effect discussed in the literature is the within-array spatial variation. Background subtraction is one approach, discussed above, for accounting for spatial variation, although systematic trends in spatial effects have been attributed to certain technologies, e.g. print tip technology (Geller et al., 2003). A generalization of (3.22) is to include an additional function $g(s(g))$ in

$$y_{gik} = \alpha_g + \beta_{gk} + f_{ik}(\alpha_g) + g(s(g)) + \epsilon_{gik} \qquad (3.24)$$

to model spatial effects, given the location map $s(g)$ of probe or gene g on the array. The function $g(s(g))$ can be very general, or in the special case of an associations with print tips, can be regarded as linear,

$$g(s(\cdot)) = Z\theta \qquad (3.25)$$

where Z is an $N \times P$ association matrix, with 1s in the respective rows of column p for genes printed with print tip p and 0s otherwise. The vector θ is assumed to be multivariate normal, with covariation matrix Ω. In this way, random associations are modeled between gene measurements by print tip blocks.

3.8   Summary and Further Reading

Our aim here is to demonstrate how simple ANOVA linear models can be extended in a straightforward manner to deal with gene expression data via hierarchical modeling. The resulting class of models is rich and leads to a coherent inferential framework for the basic scientific question being addressed. We have outlined just some of a host of other methods available (from both a Bayesian and a frequentist perspective) for gene selection and differential expression. Recent Bayesian developments include the probability of expression (POE) methods of Parmigiani et al. (2002) and stochastic regularization (West et al., 2001). Parimigiani et al. (2003) also provide a review of some recent methods, both Bayesian and frequentist, along with the associated software. For a review of some recent Bayesian advancements in microarray literature, see Do et al. (2006).

4

Bayesian Multiple Testing and False Discovery Rate Analysis

4.1 Introduction to Multiple Testing

Univariate gene detection in microarray experiments involves choosing a list of gene candidates exhibiting differential expression across known experimental conditions, e.g. over time following treatment or between normal and disease patients, see Chapter 3 for details. Consider a gene expression microarray experiment with, for the sake of simplicity, $K = 2$ experimental conditions. The univariate hypotheses are specified as

$$H_0^{(g)} : \text{Gene } g \text{ is differentially expressed,}$$

$$H_1^{(g)} : \text{Gene } g \text{ is not differentially expressed,} \qquad (4.1)$$

across factors $k = 1, \ldots, K$ for $g = 1, \ldots, G$ genes, where G is typically quite large. Suppose also that the arrays are suitably preprocessed to account for any technological variation and that each gene is represented by one and only one probe on the array. While many microarray technologies do include multiple probes to represent at least some genes, we adopt this simplifying assumption for the present. Test statistics t_g are computed for genes $g = 1, \ldots, G$ from expression data, to infer differential expression between factors. Embodied in the statistic t_g is a measure of the likelihood or in a Bayesian analysis the posterior belief that gene g is differentially expressed. A rule $\mathcal{D} : t_g \to a_g$ is specified to determine for each gene g the action a_g chosen, to accept or reject $H_0^{(g)}$. The actions $(a_1, a_2, a_3, \ldots)$ have consequences that include further experimental and

Bayesian Analysis of Gene Expression Data   B. Mallick, D. Gold, and V. Baladandayuthapani
© 2009 John Wiley & Sons, Ltd

financial costs. The enormity of the number G of univariate hypothesis tests to be performed requires attention to the practical and relevant costs and considerations for assembling a reliable collection of gene candidates in gene screening studies. This leads to us to the very important statistical consideration of *multiple testing*.

Example 4.1
Suppose that a microarray experiment is performed, with an expression array including $G = 20\,000$ genes. Suppose that independent univariate tests are performed to infer differential expression across conditions, resulting in independent frequentist p-values $(p_1, \ldots, p_G)$. For each gene one of four possible events shown in Table 4.1 occurs.

Table 4.1   Inference for differential expression

	No differential expression	Differential expression
Fail to reject H_0	Correct decision	Type II error
Reject H_0	Type I error	Correct decision

Setting a p-value cutoff of $\alpha = 0.05$ is expected to produce $0.05 \times 20\,000 = 1000$ Type I errors, or *false positive* rejections. In this special case, the standard deviation of the false positive count is $\sqrt{20\,000 \times 0.05 \times (1 - 0.05)} \approx 31$. That is, even if no genes actually exhibit differential expression, and the test assumptions are valid, the 0.05 p-value cutoff is likely to yield between approximately 938 and 1062 false positives. In such a situation as this, further time and resources would have to be allocated to follow-up study on all of the important gene candidates, to better understand their biological roles in the experiment at hand. Suppose that after considering the time and cost, the principal investigator decides to allocate resources to follow up on only 50 gene candidates. Moreover, the investigator asks you to determine which gene candidates you recommend for further follow-up.

In the above example, the time spent on follow-up study of 1000 false positive gene candidates is prohibitive. Not only would this be an inefficient expense in terms of time and money, but also it would divert resources away from other potential experimental research. Conservative tests, on the other hand, might miss important gene bio-markers. This begs the question: what cutoff should be used to decide which gene candidates are important? In general, the goals of experiments can vary depending on the objectives and costs. Balancing the relative costs of false and missed discoveries is an important consideration that can be included in a decision-theoretic framework. Many approaches have been proposed, with very different theoretical properties. In Section 4.2 we discuss false discovery rate analysis, and extensions from a frequentist point of view. In Sections 4.3 and 4.4 we turn to Bayesian FDR analysis for control of FDR in gene expression studies. A discussion of the decision-theoretic framework and of posterior and predictive inferences is given in Appendix A.1.6

4.2   False Discovery Rate Analysis

4.2.1   Theoretical Developments

In gene expression studies, a popular decision rule is based in principle on controlling the *false discovery rate*. The FDR is generally defined to be the expected percentage of positive tests that are incorrectly rejected. In gene detection analysis FDR is the expected proportion of genes that are detected for change, that are in fact not differentially expressed. An important complication with this definition is that the true number of differentially expressed genes is unknown, requiring innovative theoretical work. Before we review the relevant frequentist work on FDR estimation, let us begin with the earliest work in multiple comparison, dating back to Bonferroni. Bonferroni proposed that, rather than controlling the probability of a Type I error for each individual test, as in Example 4.1, one control the family-wise error rate (FWER) defined as

$$FWER = P(\text{'at least one false positive'}).$$

This is equal to $1 - P(\text{'no false positives'})$. If the tests are independent, and controlled at the same level α, then the $P(\text{'no false positives'}) = (1 - \alpha)^G$ and

$$FWER = 1 - (1 - \alpha)^G, \tag{4.2}$$

where G is the number of genes tested. The *Bonferroni method* controls the FWER by setting the level of significance for each test equal to

$$\tilde{\alpha} = 1 - (1 - \alpha)^{1/G} \approx \frac{\alpha}{G}.$$

The Bonferroni method ensures that the probability of at least one Type I error is less than α. For example, if $G = 20\,000$ genes are tested for differential expression, and the desired significance level is $\alpha = 0.05$, then tests with a p-value less than or equal to $\tilde{\alpha} = 0.05/20,000 = 2.5 \times 10^{-6}$ are rejected. This rule is considered to be conservative, and in some cases too stringent. Many genes that are differentially expressed at low to moderate levels can be missed with this small threshold, sacrificing power. There are several important advantages to working with the FDR rather than the FWER. Frequentist p-value thresholds convey the probability that a test will be rejected given that the null is true. The FDR is a better-understood and more intuitive concept for multiple testing, conveying the probability that a test is incorrectly rejected. A 10% FDR is consistent with the expectation that for every 10 gene detected, only one will be a false positive, a 20% FDR with 2 in 10 false positive genes, etc. Rather than controlling the

FWER, which can greatly impede detection, an investigator may be willing to accept some false positives, in order to discover important biomarkers. In theory, FDR analysis provides the investigator with a clear intuition for choosing the FDR threshold.

Consider the 2×2 table of counts of outcomes for m tests, shown in Table 4.2.1. For each test, a test positive is a test for which the null hypothesis is rejected, a test negative is a test for which the null hypothesis is not rejected, and a false positive is test for which the null is rejected when in fact the null is true, i.e. a Type I error. The number of false positives in Table 4.2.1 is V and the proportion of false positives is V/R, for $R > 0$, and 0 if $R = 0$. The FDR is the expected value of this ratio given $R > 0$. In practice V is unknown. In the special case of a binomial action, either action a_0 or a_1 is chosen. The frequentist decision rule minimizes the risk of making a wrong decision. In the context of FDR, there are $g = 1, \ldots, G$ actions sought, and the decision rule minimizes the risk or, in terms of actions, expected number of false positive and false negative decisions.

Table 4.2   Multiple comparison counts, test outcomes by truth

	Accept null	Reject null	Total
Null true	U	V	m_0
Alternative true	T	S	m_1
Total	W	R	m

Benjamini and Hochberg (1995) defined the FDR as

$$FDR = E\left[\frac{V}{R} | R > 0\right] P(R > 0), \tag{4.3}$$

the expected value of the ratio of false positives to the number of positive tests, multiplied by the probability that a test is rejected. For m tests, with m ordered p-values, $p_{(1)} \leq p_{(2)} \leq \cdots \leq p_{(m)}$, they reject all hypotheses with a p-value $p \leq p_{(\hat{j})}$, where

$$\hat{j} = \text{argmax}_{1 \leq j \leq m} k : p_{(j)} \leq \alpha \cdot j/m.$$

This yields

$$FDR = \frac{m_0}{m} \cdot \alpha \leq \alpha.$$

Benjamini and Hochberg show that choosing the p-value threshold in this way is asymptotically equivalent to choosing the threshold at the point $p*$ where the lines $\beta \cdot p$ and $F(p)$ cross, for

$$\beta = \frac{1/\alpha - \pi_0}{1 - \pi_0},$$

where π_0 is the probability that H_0 is true and $F(p)$ is the CDF of all p-values. Among all 'last crossing' rules, Benjamini and Hochberg (1995) show that their rule is asymptotically optimal, in the sense of minimizing the risk of incorrectly classifying the genes as differentially or nondifferentially expressed (Genovese and Wasserman, 2002). Benjamini and Hochberg's (1995) FDR is not the optimal Bayes rule, although it does approach the Bayes rule for large G and large true differences.

Like Bonferroni, this approach can suffer from a loss of power. Benjamini and Hochberg (1995) also offered a less conservative estimator of the FDR by replacing k/m with $k/\hat{m}$, where $\hat{m} = m\hat{\pi}_0$ is an estimator of $m\pi_0$, the number of true negatives.

Storey (2003) discusses the relative advantages of what he calls the positive FDR (pFDR), defined as

$$pFDR = E\left[\frac{V}{R} | R > 0\right],$$

omitting $P(R > 0)$ in (4.3), arguing that the case where $R = 0$ is generally not of interest. Storey (2002) proposed the FDR estimator

$$\widehat{FDR} = \frac{\hat{\pi}_0 \alpha}{\max\{R(\alpha), 1\}}, \tag{4.4}$$

where $\hat{\pi}_0$ is an estimator of π_0, the proportion of true negatives, and $R(\alpha)$ equals the number of p-values less than or equal to α. Storey related the FDR to the posterior probability of H_0. Consider rejecting H_0 if the t-test statistic t is in the rejection region Γ. The pFDR may be expressed as

$$
\begin{aligned}
pFDR(\Gamma) &= P(H_0 \text{ is True} | t \in \Gamma) \\
&= \frac{\pi_0 \cdot P(\text{Type I error of } \Gamma)}{\pi_0 \cdot P(\text{Type I error of } \Gamma) + \pi_1 \cdot P(\text{Power of } \Gamma)}
\end{aligned} \tag{4.5}
$$

for the test statistic t and test significance region Γ, where π_0 and π_1 are the prior probabilities of accepting/rejecting the null respectively (see Storey, 2003), assuming the tests are identical. Storey (2003) defines what he calls the q-value, which he argues is the Bayesian analog of a p-value, as

$$q(t) = \inf_{\Gamma_\alpha : t \in \Gamma_\alpha} pFDR(\Gamma_\alpha),$$

the smallest value of the pFDR consistent with a test statistic in the rejection region Γ_α. This relation holds for independent and weakly dependent hypothesis tests. The q-value is not the pFDR, but rather the posterior probability of H_0, given that the data t is in the minimal rejection region which includes t. While Storey offers a Bayesian interpretation of the q-value, it does not have an analog in Bayesian inference. The concept of power for detecting t in Γ is not a natural one in a Bayesian setting, where by definition inference is defined on the parameter space given the data. Chi (2007) considered the importance of sample size in determining power with Storey's pFDR.

Bickel (2004) introduced the decisive FDR (dFDR), defined as

$$dFDR(\Gamma) = \frac{E_\Gamma \left(\sum_{i=1}^m V_i \right)}{E_\Gamma \left(\sum_{i=1}^m R_i \right)}, \tag{4.6}$$

considering its useful decision-theoretic properties. The desirability function, given the benefit, b_g, and cost, c_g, of rejecting the gth hypothesis, is

$$d(b, c) = \sum_{g=1}^m (b_i (R_i - V_i) - c_i V_i) = \sum_{g=1}^m b_i R_i - \sum_{g=1}^m (b_i + c_i) V_i. \tag{4.7}$$

the expected desirability, given b_g and c_g, is

$$Ed(b, c, \Gamma) = b_1 \left(1 - (1 + c_1/b_1)dFDR(\Gamma) \right) E \left(\sum_{g=1}^m R_i \right). \tag{4.8}$$

A rejection region can be chosen such that a given expected desirability is achieved. Moreover, Bickel (2004) explained that the dFDR more directly approximates the posterior probability of the null hypothesis than the pFDR.

For those interested in advanced topics, Genovese and Wasserman (2004) developed a theoretical framework for construction of frequentist false discovery proportion (FDP) confidence envelopes. Muller et al. (2004) developed optimal sample size criteria for bivariate loss functions, including a criterion for both the FDR and the false negative rate (FNR), i.e. the probability of missing a true positive. They showed that the FNR decays at rate $O\sqrt{\log n/n}$ where n is the number of samples, i.e. subjects, in the study. Yekutieli and Benjamini (2001) propose a resampling-based approach to FDR analysis with correlated tests. Storey (2002) discusses weak and positive dependence. Bickel (2004) argues that the dFDR holds regardless of assumptions about independence or weak dependence between the tests. The assumption of weak dependence is difficult to justify in practice. It is generally believed that gene-wise expression is dependent. Much of the development in univariate gene detection includes the assumption of gene-wise independence, for practical reasons: the uncertainty in gene expression, the uncertainty in gene networks, as well as the intractable mathematics involved in imposing the assumption of gene-wise dependence.

4.2.2   FDR Analysis with Gene Expression Arrays

Methodological developments in frequentist FDR analysis for gene detection extend the theoretical testing framework by pooling information between test statistics with a broad array of non-parametric and bootstrap resampling techniques aimed at robustness (see Westfall and Young, 1993). Efron and Tibshirani (2002) proposed what they called an empirical Bayes method for gene detection. Efron discussed the relevance of properly modeling both the null and alternative distributions of the test statistics, using robust methods. He cited, among others, the importance of properly accounting for technological variation possibly due to gene-wise correlation or unexplained technological variability inducing gene-wise dependence, and the consequences for FDR analyses that pool information from univariate tests. Suppose that the change in gene expression, e.g. between cancer and normal, is summarized by a z-score, $\{z_g; g = 1, \ldots, G\}$, the difference in group averages divided by an estimate of the standard deviation. Efron applies an empirical quantile transformation to the t-statistics, in order to obtain standard normal z-scores

$$z_g = \Phi^{-1}(EDIFY(t_g))$$

where *EDIFY* is the empirical CDF of the test statistics. Efron models the z-scores as a mixture of random variables,

$$P(z_g) = \pi_0 P_0(z_g) + \pi_1 P_1(z_g),$$

where $P_0(z)$ is assumed to be the distribution of z if H_0 is true, and $P_1(z)$ if H_1 is true. Efron deliberated on the problem of robust estimation of the nonnull distribution of the tests statistics. While $P(z)$ is estimated directly from the data, $P_0(z)$ is estimated on a permuted version of the data, with sign permutations on differences between observations within each respective group. The probability of no change, π_0, is estimated given $\hat{P}(z)$ and $\hat{P}_0(z)$. Efron's *local false discovery rate* (lFDR) is

$$lFDR(z) = \pi_0 P_0(z)/P(z). \tag{4.9}$$

Efron relates the lFDR to the FDR as follows:

$$FDR(\Gamma) = P(\text{null}|z \in \Gamma) = \int_\Gamma lFDR(z)P(z)dz. \tag{4.10}$$

Pounds and Morris (2003) also model the p-value distribution as a mixture. In their original work, the p-values were modeled as a beta-uniform mixture (BUM),

$$P(p) = \pi_0 P_0(p) + (1 - \pi_0)P_1(p),$$

with $P_0(p)$ the uniform part and $P_1(p)$ the nonuniform part. The nonuniform part P_1 is assumed to be a beta distribution with parameters a and b estimated from the data. The motivation for this method is that, under strict conditions, the theoretical distribution of p-values for which the null hypothesis is true is a uniform distribution, and for tests for which the alternative is true, the theoretical distribution is beta with expectation less than 0.5. In practice, it has been well documented that the nonuniform part of the p-value distribution does not follow a beta distribution, and can exhibit signs of multimodality. Pounds and Cheng (2006) extended the BUM to model the nonuniform component of the p-value distribution $P_1(p)$ by nonparametric regression. Another method is provided by Dahl and Newton (2007) who conduct multiple hypothesis testing by clustering treatment effects.

Example 4.2
We return to the Wachi et al. (2005) data set consisting of 22 283 expression probe values from ten patients, five with SCC and five normal lung patients (see Section 3.6). For each gene, the hypothesis test was performed comparing the mean log-ratio δ_g in gene expression between the groups: $H_0^{(g)} : \delta_g = 0$ versus $H_1^{(g)} : \delta_g \neq 0$, assuming independent lognormal residuals. The BUM results are plotted in Figure 4.1. While a 10% FDR cutoff is usually considered conservative, the present analysis results in 6714 probe candidates with approximately 670 expected false positives!

Allison et al. (2002) extended the p-value mixture model in a purely parametric framework, treating the distribution of the p-values as mixtures of beta random variables

$$P(p) = \sum_{j=1}^{J} \pi_j \text{Beta}(p; a_j, b_j) \tag{4.11}$$

with parameters $a_1 = 1$ and $b_1 = 1$ in the first component. Given a total of J components, sensible order constraints can be placed on the a_j and b_j, in order to assure identifiability. The log-likelihood with $J*$ components is

$$L_{j*}(\mathbf{p}) = \sum_{i=1}^{G} \log \left[\sum_{j=1}^{J*} \pi_j \text{Beta}(p_g; a_j, b_j) \right].$$

Allison applies a bootstrap approach to choosing the number of components with the statistic $Q = 2(L_{j*}(p) - L_{j*-1}(p))$. Consider the mixture model of Allison (4.11) with FDR estimator (4.12)

$$P(H_0|p \leq \alpha) = \frac{P(H_0 \cap p \leq \alpha)}{P(p \leq \alpha)}$$

where

$$P(p \leq \alpha) = \tau_1 \alpha + \sum_{j=2}^{J} \int_0^{\alpha} \pi_j \text{Beta}(p; a_j, b_j) dp$$

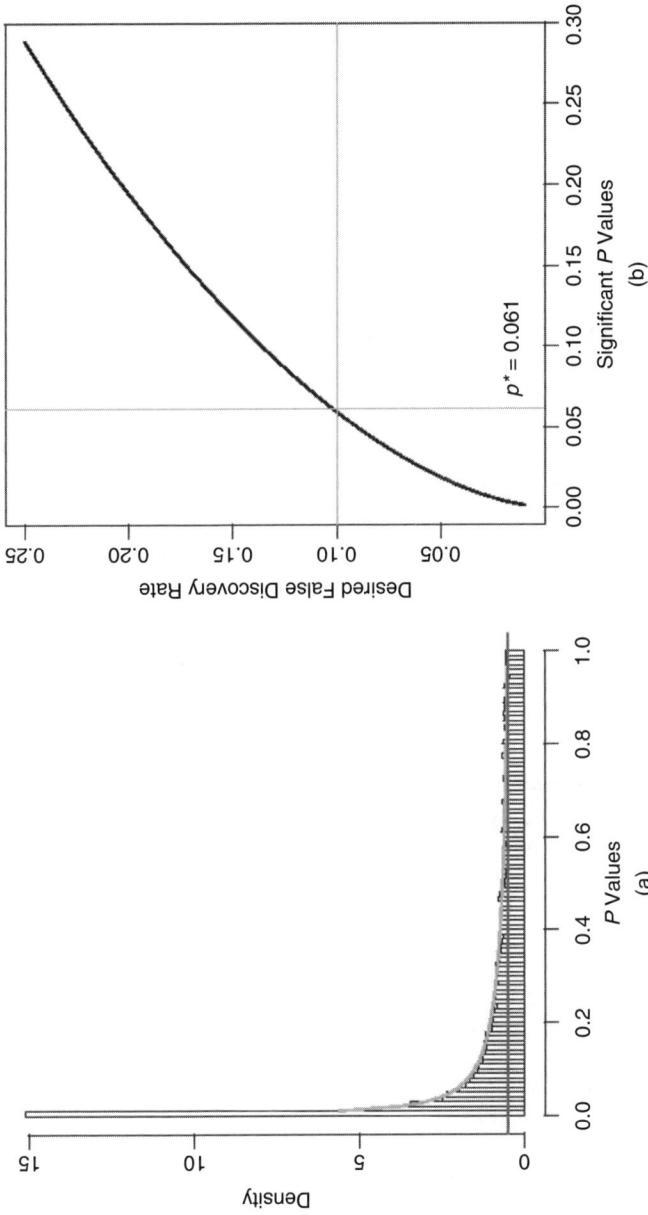

Figure 4.1 BUM analysis. (a) p-value distribution with fitted Beta mixture density. (b) BUM FDR estimates: a 10% FDR cutoff corresponds to a 0.061 p-value cutoff, resulting in 6714 probe candidates.

and

$$P(H_0 \cap p \leq \alpha) = \pi_1 \alpha,$$

so that

$$P(H_0|p \leq \alpha) = \frac{\pi_1 \alpha}{\pi_1 \alpha + \sum_{j=2}^{J} \int_0^\alpha \tau_j \mathrm{Beta}(p; a_j, b_j) dp}. \tag{4.12}$$

This is Allison's estimator of FDR. They propose their bootstrap approach for calculating confidence intervals around $\widehat{FDR}$.

4.3 Bayesian False Discovery Rate Analysis

4.3.1 Theoretical Developments

Suppose that gene expression measurements are conditionally independent, $x_g \sim P(\theta_g)$, for genes $g = 1, \ldots, G$, with the prior for θ_g denoted by $P(\theta_g)$. Suppose further that the parameter space of $\theta_g \in \Theta$ can partitioned into two sets $\Theta = (\Theta_0, \Theta_1)$ with the null hypothesis for gene g defined as $H_0^{(g)} : \theta_g \in \Theta_0$ and the alternative $H_1^{(g)} : \theta_g \in \Theta_1$. The Bayesian decision rule rejects $H_0^{(g)}$ if

$$P(\theta_g \in \Theta_1 \mid x_g) > (1 - \alpha), \tag{4.13}$$

for some user-specified α. In a Bayesian framework the FDR is a measure of the posterior belief of the event $\theta_g \in \Theta_0$ given that $H_1^{(g)}$ was chosen, i.e. gene g was discovered to be biologically important. Unlike frequentist testing, $H_0^{(g)}$ is assumed unknown according to a prior belief that is updated from data. Let the variable $r_{\Theta_1}^\alpha(x_g) = 1$ if $H_1^{(g)}$ is chosen and $r_{\Theta_1}^\alpha(x_g) = 0$ otherwise. The Bayesian FDR (bFDR) for a single gene g is defined as $P(H_0^{(g)}|r_{\Theta_1}^\alpha(x_g), x_g)$. In an experiment with G genes, the bFDR is defined as

$$bFDR(r_{\Theta_1}^\alpha) = \frac{\sum_g P(\theta_g \in \Theta_0 \mid r_{\Theta_1}^\alpha(x_g), x_g) \cdot r_{\Theta_1}^\alpha(x_g)}{\sum_g r_{\Theta_1}^\alpha(x_g)}, \tag{4.14}$$

(Muller et al., 2004; Whittemore, 2007). In like manner, the Bayesian true negative rate (bTNR) is defined as

$$bTNR(r_{\Theta_1}^\alpha) = \frac{\sum_g P(\theta_g \in \Theta_0 \mid r_{\Theta_1}^\alpha(x_g) = 0, x_g) \cdot (1 - r_{\Theta_1}^\alpha(x_g))}{\sum_g (1 - r_{\Theta_1}^\alpha(x_g))}. \tag{4.15}$$

Notice that, like Storey's FDR, the bFDR depends on the rejection decision rule and consequently on (Θ_1, α). In practice, the set Θ_1 is fixed, and the decision rule is allowed to vary with the choice of α. The local bFDR for gene g, $lbFDR_g$, is defined as

$$lbFDR_g = bFDR(r_{\Theta_1}^{\alpha*}), \quad \text{for } \alpha* = \text{argmin}_\alpha r_{\Theta_1}^{\alpha}(x_g) = 1, \qquad (4.16)$$

which may be reported for each gene.

Multivariate generalizations of the bFDR follow from the definition. The null hypothesis for gene set $\mathcal{G} = \{g_1, g_2, g_3, \ldots\}$ is defined as $H_0^{(\mathcal{G})} : \theta_\mathcal{G} \in \Theta_0^{(\mathcal{G})}$ and the alternative as $H_1^{(\mathcal{G})} : \theta_\mathcal{G} \in \Theta_1^{(\mathcal{G})}$. The posterior rejection region is

$$P(\theta_\mathcal{G} \in \Theta_1^{\mathcal{G}} \mid x) > (1 - \alpha(|\mathcal{G}|)), \qquad (4.17)$$

for user-specified $\alpha(|\mathcal{G}|)$, a function of the cardinality of the set $\mathcal{G}$. Let the variable $r_{\Theta_1^{(\mathcal{G})}}^{\alpha(|\mathcal{G}|)}(x) = 1$ if $H_1^{(\mathcal{G})}$ is chosen and $r_{\Theta_1^{\mathcal{G}}}^{\alpha(|\mathcal{G}|)}(x) = 0$ otherwise. In order to estimate the multivariate $bFDR(r_{\Theta_1^{(\mathcal{G})}}^{(\alpha(|\mathcal{G}|))}(x))$, a measure is needed of the posterior probability $P(\theta_\mathcal{G} \in \Theta_0^{(\mathcal{G})}|x)$, in the case where $H_0^{(\mathcal{G})}$ is rejected. The Bayesian multivariate false discovery rate (bMVFDR) is defined as this conditional posterior probability,

$$bMVFDR(r_{\Theta_1^{(\mathcal{G})}}^{(\alpha(|\mathcal{G}|))}(x)) = P(H_0^{(\mathcal{G})}|r_{\Theta_1^{(\mathcal{G})}}^{(\alpha(|\mathcal{G}|))}(x) = 1, x). \qquad (4.18)$$

A practical consequence of this definition is a framework for the Bayesian estimation of the FDR of gene or probe combinations. For instance, consider a microarray design with replicate probes for at least some of the genes. Prior information about replicate probes, e.g. dependence, relative binding efficiency or signal strength, can be specified flexibly through the prior in order to form bFDR estimates pooling across the probes mapping to each gene.

4.4 Bayesian Estimation of FDR

Bayesian extensions of frequentist parametric FDR inference are straightforward. The essential difference is that in a Bayesian hypothesis test, rather than estimating FDR with plug-in estimators for model parameters, a Bayesian FDR analysis is conducted by integrating over all of the posterior uncertainty in the parameter space. Consider the mixture model of Allison (4.12) with FDR estimator (4.13). The Bayesian FDR is estimated by

$$\cdot bFDR(p < \alpha) = \int \cdots \int \psi(p < \alpha|\pi, \mathbf{a}, \mathbf{b}) P(\pi, \mathbf{a}, \mathbf{b}|\mathbf{p}) d\pi d\mathbf{a} d\mathbf{b} \qquad (4.19)$$

with

$$\psi(p < \alpha|\pi, \mathbf{a}, \mathbf{b}) = \frac{\pi_1 \alpha}{\pi_1 \alpha + \Sigma_{j=2}^J \int_0^\alpha \pi_j \text{Beta}(p; a_j, b_j) dp} \qquad (4.20)$$

integrating over the posterior uncertainty in π, $\mathbf{a}$, $\mathbf{b}$. Bayesian posterior credible envelopes for the quantity $\psi(p < \alpha | \pi, \mathbf{a}, \mathbf{b})$ over the range $(0, 1)$ can be computed in a straightforward manner as well following the method outlined for determining credible envelopes in Chapter 3.

Next, we demonstrate the flexibility of the Bayesian decision-theoretic approach, not to be confused with Bayesian hypothesis testing, generalizing the FDR analysis in Example 13.1 with a continuous null hypothesis region.

Example 4.3
Returning to the analysis in Example 13.1, suppose that we pose the hypotheses as $H_0^{(g)} : |\delta_g| \leq 1$ versus $H_1^{(g)} : |\delta_g| > 1$, where δ_g is the mean log-ratio of gene expression between cancer and normal in the Wachi experiment for gene g. Following the assumptions in Example 13.1, the likelihood of

$$\frac{\bar{x}_{g1} - \bar{x}_{g2} - \delta}{\hat{\sigma}_g \sqrt{(1/5 + 1/5)}} \sim t_8, \tag{4.21}$$

for $g = 1, \ldots, G$, difference in group means $\bar{x}_{g1} - \bar{x}_{g2}$, and standard deviation estimate $\hat{\sigma}_g$, follows a t distribution with 8 degrees of freedom. Specifying an improper flat prior for δ_g results in the posterior distribution

$$\delta_g | x_g \sim t_8(\bar{x}_{g1} - \bar{x}_{g2}, \hat{\sigma}_g^2 (1/5 + 1/5)) \tag{4.22}$$

a t with 8 degrees of freedom, location parameter $\bar{x}_{g1} - \bar{x}_{g2}$, and dispersion

$$\hat{\sigma}_g^2 \left(\frac{1}{5} + \frac{1}{5} \right). \tag{4.23}$$

The Bayesian decision rule rejects $H_0^{(g)}$ if

$$P(\delta_g < -1 | x_g) + P(\delta_g > 1 | x_g) > (1 - \alpha) \tag{4.24}$$

for user-specified α. Different choices of α lead to different decisions, more or less rejected hypotheses, and hence different bFDRs. The bFDR in this example is

$$bFDR(r^\alpha) = \frac{\sum_g P(-1 < \delta_g < 1 | x_g) r^\alpha(x_g)}{\sum_g r^\alpha(x_g)}, \tag{4.25}$$

depending on α, where $r^\alpha(x_g) = 1$ if $H_0^{(g)}$ is rejected and 0 otherwise. The corresponding likelihood ratio frequentist test of $H_0^{(g)} : |\delta_g| \leq 1$ versus $H_1^{(g)} : |\delta_g| > 1$ yields 21 431 genes with p-values of 1, and few p-values between 0 and 1, i.e. strong evidence against a continuous p-value distribution. Recall also that Efron's approach to modeling change in gene expression relies on permuted versions of the data, i.e. differences between and within the normal and cancer groups respectively, and therefore does not permit inference against a continuous null set such as $H_0^{(g)} : |\delta_g| \leq 1$.

Robust innovations for FDR estimation combined with Bayesian methods have been proposed for multiple testing with gene expression data analysis. Ishwaran and Rao (see BAM in Appendix A), propose the strategy of simulating data Y_{new} from the posterior predictive distribution assuming H_0 is true to estimate a null distribution, and consequently the FDR. Recall that the transformed response in BAM is

$$\tilde{Y}_{gkl} = \theta_{g0} + \mu_{gl} I (l = 2) + \epsilon_{gkl}, \qquad (4.26)$$

with ϵ_{jkl} assumed to be residual noise. Due to the centering $Y_{gkl} - \bar{Y}_{gl}$, θ_{g0} is expected to be near 0. The posterior distributions are known in the case where $\mu_{gl} = 0$ and gene g is selected for change, i.e. v_{2g}^2 is large. The statistic

$$\frac{\sqrt{n_2}}{\hat{\sigma}_n} (\bar{Y}_{g1} - \bar{Y}_{gx2}) \qquad (4.27)$$

is normally distributed with variance $(n_1 + n_2)/n_1$. Ishwaran and Rao fit a two-point mixture model to data simulated from their model, what they call FDRmix, constraining $\mu_{jl} = 0$. The simulated data provides estimation of a null distribution for each gene, and a p-value for change. Their FDR is controlled as proposed by Benjamini and Hochberg. This is a two-stage analysis, making use of the posterior estimates in the first stage, to model the FDRs for the parameters of change in the second stage.

Do et al. (2005) applied robust mixture modeling to permuted versions of the data to estimate FDR. In the case where microarray data are collected from two groups, disease and normal phenotypes, a matrix D is generated with, for each gene (in the rows), all possible pairs of differences between the tumor and normal samples (in the columns). A matrix d is constructed similarly, with differences between all possible pairs of samples within the tumor and normal groups respectively. For genes $g = 1, \ldots, G$, two z statistics are generated

$$z_g^{\text{mix}} = \bar{d}_g'/(a_0' + \hat{\sigma}_g'),$$
$$z_g^{\text{null}} = \bar{d}_g/(a_0' + \hat{\sigma}_g),$$

where z_g^{mix} and z_g^{null} are standardized Z statistics computed across the columns of d' and d, with sample standard deviations $\hat{\sigma}_g'$ and $\hat{\sigma}_g$. The constants a_0' and a_0 are added for robustness. Extending Efron's work, Do et al. model z^{mix} as a mixture

$$P(Z) = \pi_0 P_0(z) + \pi_1 P_1(z)$$

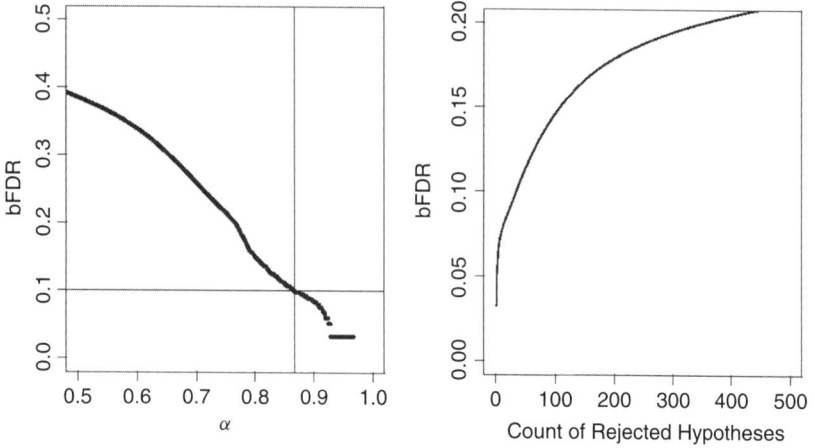

Figure 4.2   Bayesian FDR analysis in Example 13.2 (a) bFDR by decision rule cutoff α, (b) bFDR by count of rejected hypotheses, choosing bFDR = 10% results in 33 rejected hypotheses.

with likelihood for z^{null} assumed to be $P_0(z)$. For $j = 0, 1$,

$$P_j(z) = \int N(z; \mu, \sigma^2) d\Gamma_j(\mu),$$

$$\Gamma_j \sim DP(a, \Gamma_j^*).$$

The means are assumed to follow a Dirichlet process prior $DP(a, \Gamma_j^*)$, with scale a and base measure Γ_j^*. Do et al. define FDR by (14), applying the rejection rule $EP_1(z^*) = E(1 - \pi_0)P_1(z^*)/P(z^*) > 1 - \alpha$, with the expectation taken over the posterior distribution of the parameters and α specified as the smallest cutoff achieving a desired FDR.

Example 4.4

We conclude our discussion of Bayesian FDR analysis with application of Do et al.'s robust Dirichlet process mixture model to the Wachi et al. (2005) data set. Information about obtaining the BayesMix software is available from the authors' supplementary materials. BayesMix was run with correction factors $a_0 = a_0' = 0.50$. Figure 4.3 shows the densities of the empirical scores z^{null} and z^{mix}. The tails of the density of the latter are considerably heavier than the former. Output from BayesMix is displayed in Figure 4.4. At the 0.10 FDR 12 664 probes were discovered for change, at the 0.05 FDR 8487, and at the 0.01 FDR 5404 probes. The DP mixture model is quite robust, relying on very little in the way of assumptions about the underlying distributions. Unlike other software, BayesMix is quick and easy to apply, and relies on few user-controlled parameters.

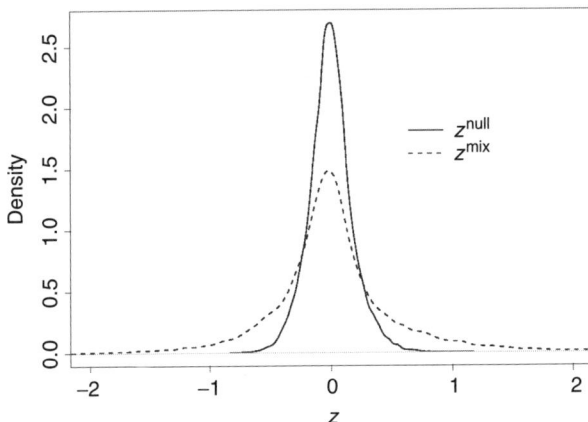

Figure 4.3   Kernel density plots z^{null} and z^{mix} with Wachi et al. (2005) experimental gene expression data.

4.5   FDR and Decision Theory

The multiple testing problem was posed in a decision-theoretic framework by Duncan (1965), who laid the groundwork for Bayesian analysis of multiple comparisons. His early work surpassed even much later thinking, that the problem of choosing a rejection rule can be driven by competing goals. Conventional thinking in multiple testing recognizes the importance of the decision-theoretic framework; see; for example; Berry and Hochberg (2001) and Chen and Sarkar (2005). Working with the posterior of the FDR function offers theoretical simplicity for Bayesian analysis. Analogously, the Bayesian false negative rate (bFNR) function can be defined as

$$bFNR(r^\alpha_{\Theta_1}) = \frac{\sum_g P(\theta_g \in \Theta_1 \mid r^\alpha_{\Theta_1}(x_g), x_g)) \cdot (1 - r^\alpha_{\Theta_1}(x_g))}{\sum_g (1 - r^\alpha_{\Theta_1}(x_g))}, \qquad (4.28)$$

In a decision-theoretic setting, the choice of the best threshold is guided by the relative costs of misspecification. Some popular loss functions include linear loss $FDR + \omega FNR$ and bivariate loss $FNR|FDR < \alpha$. Posterior evaluation of the loss yields probability measures over the full posterior uncertainty in decision making. The posterior uncertainty in decisions can be explored in silico, to inform choices of study design; see Muller et al. (2004) for extensive discussion.

4.6   FDR and bFDR Summary

Interest in FDR analysis has grown considerably with the advent of new array technologies, and the accumulation of ever increasing computing storage capacities. These advances have provoked researchers to expand the definitions of

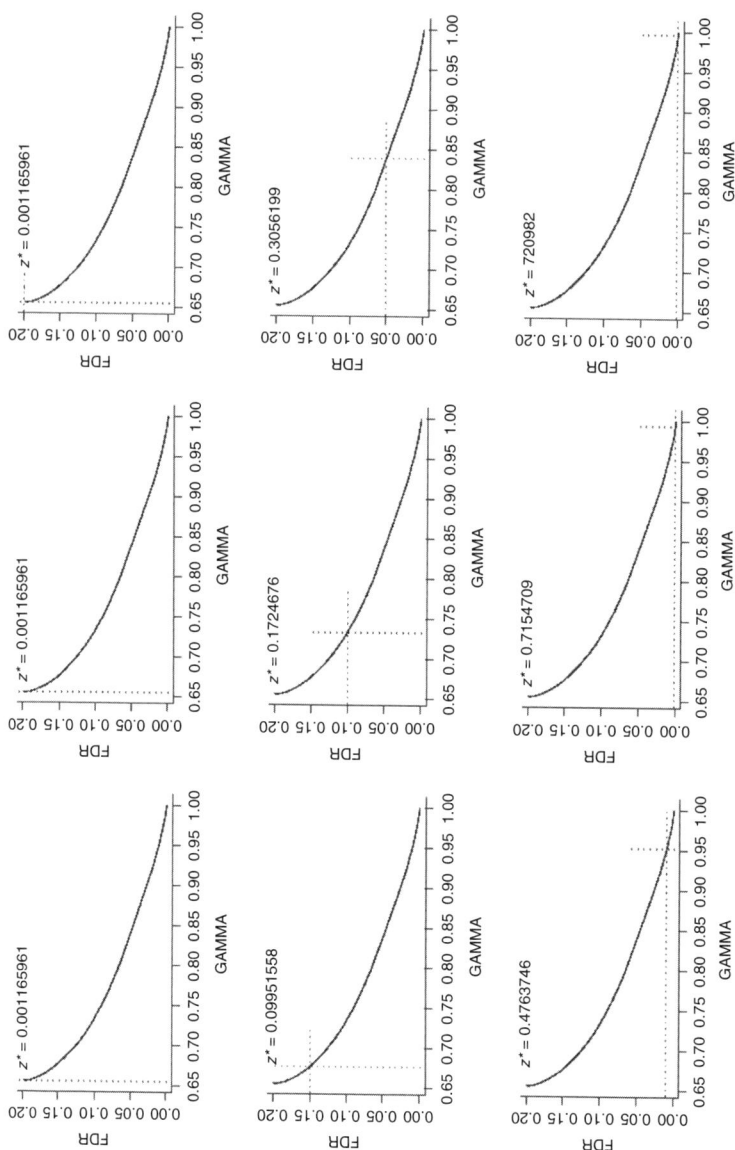

Figure 4.4 BayesMix, applied to Wachi et al. (2005) experimental gene expression data. z-value cutoffs are shown for different desired FDRs.

what it means to infer that a set of variables are important for discovery. While there is little consensus on how to estimate or even define the FDR, FDR analyses provide a practical rule for the problem of multiple testing with microarray data. Further statistical considerations include addressing more complicated experimental designs, and combining multiple per gene hypotheses, of many different characteristics measured across a variety of sources. Other important topics, for example FDR confident envelop estimation (Genovese and Wasserman, 2004; Allison et al., 2002) or sample size requirements (Muller et al., 2004; Chi, 2007), are still very much active areas of research.

Much of the underlying theory of FDR analysis can be posed in a decision-theory framework, and this framework allows contributions from both frequentist and Bayesian schools of thought. Ad hoc frequentist and Bayesian methods for FDR analysis are growing more plentiful, for a list of some of the most important advances, see Pounds and Cheng (2007). We have presented some of the ground-breaking ideas in frequentist and Bayesian FDR estimation. There is still much to be discovered. As we continue to expand our vocabulary in new ways that allow us to think about multiple testing with ever large data sets to consider, these important new ideas will filter into other arenas, providing concepts of inference to the realms medical and biological informatics.

5

Bayesian Classification for Microarray Data

5.1 Introduction

One of the fundamental goals of microarray data analysis is classification of biological samples. The problem of classification or supervised learning has received a great deal of attention, especially in the context of cancer studies. Precise classification of tumors is often of critical importance to the diagnosis and treatment of cancer. Targeting specific therapies to pathogenetically distinct types of tumor is important for the treatment of cancer because it maximizes efficacy and minimizes toxicity (Golub et al., 1999). Initial studies relied on macroscopic and microscopic histology and tumor morphology as the basis for the classification of tumors. However, within current frameworks, one cannot discriminate between tumors with similar histopathologic features, which vary in clinical course and in response to treatment.

The use of DNA microarrays allows simultaneous monitoring of the expressions of thousands of genes (Schena et al., 1995; DeRisi et al., 1997; Duggan et al., 1999), and has emerged as a tool for disease diagnosis based on molecular signatures. Several studies using microarrays to profile colon, breast and other tumors have demonstrated the potential power of expression profiling for classification (Alon et al., 1999; Golub et al., 1999; Hedenfalk et al., 2001). Gene expression profiles may offer more information than and provide an alternative to morphology-based tumor classification systems. In this chapter we focus on classification using microarray data. In principle, gene expression profiles might serve as molecular fingerprints that would allow for accurate classification of

Bayesian Analysis of Gene Expression Data   B. Mallick, D. Gold, and V. Baladandayuthapani
© 2009 John Wiley & Sons, Ltd

diseases. The underlying assumption is that samples from the same class share expression profile patterns unique to their class (Yeang et al., 2001). In addition, these molecular fingerprints might reveal newer taxonomies that previously have not been readily appreciated.

In general, the problem of class detection can be categorized based on two main aspects: *clustering* or *supervised learning*, and *classification* or *unsupervised learning*. In unsupervised learning (other nomenclatures include cluster analysis, class discovery, and unsupervised pattern recognition), the classes are unknown a priori and are estimated from the observed data. Note that the clustering mechanism can be applied to both the genes and the microarrays (samples). In contrast, in supervised learning or classification, the classes are predefined and the goal is build a classifier using the attributes of the data and predict the class of future unlabeled observations. Although there are inherent connections between the two, we shall focus on supervised learning/classification for the remainder of the chapter. See Chapter 7 for clustering techniques applied to microarrays.

In classical multivariate statistics, classification is a prediction or learning problem in which the response variable Y to be predicted assumes one of C predefined and unordered values arbitrarily coded as dummy variables $(1, 2, \ldots, C)$. For $K = 2$ this reduces to a binary classification problem with responses usually coded as $Y \in (0, 1)$, where 0 might be normal tissue and 1 tumor tissue. Also observed are a set of p measurements that form the feature/predictor vector, $\mathbf{X} = (X_1, \ldots, X_p)$ belonging the a feature space $\mathcal{X}$ (e.g. span of real numbers R^p). The task of classification is to predict Y from $\mathbf{X}$ on the basis of the observed measurements $\mathbf{X} = \mathbf{x}$. The *classifier* for the C classes is a mapping, $\mathcal{T}$, from $\mathcal{X}$ onto $(1, 2, \ldots, C)$ as $\mathcal{T} \colon \mathcal{X} \to (1, 2, \ldots, C)$, where $\mathcal{T}(\mathbf{x})$ is the predicted class for an observed feature matrix $\mathbf{X}$.

In microarrays, the features correspond to the expression values of different genes, and classes correspond (usually) to the different tumor types (normal/cancer, etc.). Gene expression data for p genes for n samples are summarized in an $n \times p$ matrix, $\mathbf{X}$, where each element X_{ij} denotes the expression level (gene expression value) of the jth gene, $j = 1, \ldots, p$, in the ith sample:

$$
\begin{bmatrix}
 & \text{Gene 1} & \text{Gene 2} & \cdots & \text{Gene } p \\
Y_1 & X_{11} & X_{12} & \cdots & X_{1p} \\
Y_2 & X_{21} & X_{22} & \cdots & X_{2p} \\
\vdots & \vdots & \vdots & \ddots & \vdots \\
Y_n & X_{n1} & X_{n2} & \cdots & X_{np}
\end{bmatrix}.
$$

The exact meaning of expression values may be different for different matrices, representing absolute or comparative measurements (see Brazma et al., 2001). In this chapter we review some Bayesian linear and nonlinear approaches to classification – where the linearity/nonlinearity assumption is made on the way the feature space is modeled. We show how the classical linear discriminant analysis can be extended in a Bayesian framework in order to account for estimation uncertainty. At the heart of our discussion of the methods lies classification via regression-based approaches. Regression methods are extremely flexible and these models lend themselves nicely to the hierarchical Bayesian modeling framework in which the variability in the gene expression data can be modeled at various levels.

One of the key issues in microarray classification as opposed to other applications is the dimension of the feature space, typically of the order of thousands of genes, coupled with a low number of arrays (usually in hundreds). This is typically known as the *large p, small n* problem (West, 2003). In this situation, in most classification procedures a dimension reduction step is needed to reduce the high-dimensional gene space. There are two main reasons for this: first, most of the genes are likely to be uninformative for prediction purposes; and second, the performance of the classifier depends highly on the number of features used (West et al., 2001; Ambroise and Mclachlan, 2002). This is termed feature selection and is a topic of extensive research in both the statistical and computer science literature. Feature selection in the context of microarrays can be done in two ways: *explicitly* and *implicitly*. In the former genes are first selected one at a time using univariate test statistics such as t- or F-statistics (Dudoit et al., 2003); ad hoc signal-to-noise statistics (Golub et al., 1999; Pomeroy et al., 2002); nonparametric Wilcoxon statistics (Park et al., 2001); and p-values (Dudoit et al,. 2002). The cutoff thresholds are then selected using some criteria such as false discovery rates. The classifier is then built upon this feature set. Such methods have a potential disadvantage since the uncertainty in estimation of the feature set is not taken into account while building and estimating the error rates of the classifier. In the latter implicit approach, the selection is embedded within the classification rule itself. This scenario can easily handled within the Bayesian hierarchical modeling framework, by specifying an extra layer for feature selection set uncertainty. Not only is the uncertainty in estimation propagated through the modeling process, but also we obtain an honest assessment of this uncertainty via posterior sampling, as we shall describe in the subsequent sections. Section 5.2 starts by exploring the classical classification and discriminant rules, and their Bayesian extensions are discussed in Section 5.3. Section 5.4 delves into regression-based approaches to classification and Section 5.5 deals with non-linear extensions. We talk about some aspects of evaluating the performance of a classifier in Section 5.6 and then round matters off with a real data example (Section 5.7) and a discussion (Section 5.8).

5.2   Classification and Discriminant Rules

Classical approaches to classification have been mainly via discriminant rules which are inherently Bayesian in some sense. For example, consider the Bayes rule under a symmetric loss function. Given class conditional densities $P_c(\mathbf{X}) =$

$P(\mathbf{X}|Y = c)$ of the features $\mathbf{X}$ in class c and class priors π_c, the posterior probability $P(Y = c|\mathbf{X})$ of class c given feature vector $\mathbf{X}$ is

$$P(Y = c|\mathbf{X}) = \frac{\pi_c P_c(\mathbf{X})}{\sum_k \pi_k P_k(\mathbf{X})}.$$

The *Bayes rule* predicts the class of an observation $\mathbf{X}$ by that with higher posterior probability $\mathcal{T}_B(\mathbf{X}) = \operatorname{argmax}_c P(Y = c|\mathbf{X})$. Under a symmetric loss function this is just the Bayes risk. We refer the reader to Mardia et al. (1979) for further discussion on classification within a decision-theoretic framework.

Many classifiers can be viewed as variations of this general rule, with parametric or nonparametric estimators of $P(Y|\mathbf{X})$ (Dudoit and Fridlyand, 2002). There are two general paradigms that are used to estimate the class posterior probabilities $P(Y|\mathbf{X})$: density estimation and direct function estimation (Friedman, 1997). In the density estimation approach, the class conditional densities $P_c(\mathbf{X}) = P(\mathbf{X}|Y = c)$, along with the class priors π_c, are estimated separately for each class and Bayes' theorem is then used to obtain $P(Y|\mathbf{X})$. Classification methods employing this principle include maximum likelihood discriminant rules and discriminant analysis. In the next section we discuss a Bayesian extension of these rules. In the direct function estimation approach, the class posterior probabilities $P(Y|\mathbf{X})$ are estimated directly based on function estimation approaches such as regression. This general framework includes most of the popular classification approaches: logistic regression, neural networks (Ripley, 1996) and classification trees (Breiman et al., 1984). We focus on this approach for most of our discussion of Bayesian regression-based methods for linear and nonlinear classification primarily due to their flexibility and popularity.

5.3   Bayesian Discriminant Analysis

Bayesian discriminant rules for classification usually proceed by extending the frequentist classification rules and assuming a prior on the parameters. To formalize, suppose that independent random variables, features or genes (possibly vectors) $X_{i1}, \ldots, X_{in_i}$ are observed from populations (classes) $i = 1, \ldots, C$, each with probability distribution $P_i(\theta_i)$, where n_i is the number of genes in class i. The likelihood can then be written as $\prod_{i=1}^{C} \prod_{j=1}^{n_i} P_i(X_{ij}|\theta_i)$ where the θ_i are the unobserved population parameters that need to estimated.

In classical frequentist classification, a new observation X_{new} is classified by estimating θ_i from the training observations, $\widehat{\theta_i}$, and plugging back into the likelihood to form the prediction rules. The new observation X_{new} is then assigned to the class i for which $P_i(X_{\text{new}}|\widehat{\theta_i}) > P_i'(X_{\text{new}}|\widehat{\theta_i'})$ for all i', and in case of ties assigned randomly. In the Bayesian paradigm, a prior is assumed over θ, in order to carry forward the uncertainty in estimation of θ while making predictions. We refer the reader to Friedman (1997) for a discussion of the bias–variance tradeoff in classification.

In Bayesian parametric classification, a new observation X_{new} is classified by assigning a prior distribution to the θ, $P(\theta_1, \ldots, \theta_C)$, and using the posterior distribution for inference. The (joint) posterior, $P(\theta_1, \ldots, \theta_C|X)$, is then proportional to $\prod_{i=1}^{C} \prod_{j=1}^{n_i} P_i(X_{ij}|\theta_i) P(\theta_1, \ldots, \theta_C)$. The predictive distribution can then be obtained by marginalizing over θ. For the ith population the predictive distribution for X_{new} can then be written as

$$P_i(X_{\text{new}}|\mathbf{X}) = \int_{\Theta_i} P_i(X_{\text{new}}|\theta_i)\pi(\theta_i|\mathbf{X})d\theta_i, \qquad (5.1)$$

for all i and where Θ_i is the support of the θ_i. The Bayesian prediction rule then assigns X_{new} to population i via the rule $P_i(X_{\text{new}}|\mathbf{X}) > P_{i'}(X_{\text{new}}|\mathbf{X})$ for all (i, i'), and in case of ties assigned randomly. This then leads to the Bayesian classification rule based on the posterior predictive distribution,

$$P(Y_{\text{new}} = i) = \frac{\pi_i P_i(X_{\text{new}}|X)}{\sum_c \pi_c P_c(X_{\text{new}}|X)}.$$

Frequentist classification methods usually rely on large-sample theory or resampling mechanisms in order the determine the uncertainty in the predictions. In the Bayesian classification rule presented above, uncertainty is quantified once the posterior $\pi(\theta_i|X)$ is obtained. This is discussed in some detail in Section 5.6 below.

In principle, one can assume any distribution for the feature space $P_i(X|\theta)$ depending on the application at hand. Assuming a Gaussian distribution gives rise to the most common class of Bayesian linear classifiers, which are linear on the feature space. Suppose the feature vectors $X_{i1}, \ldots, X_{in_i}$ are independently observed from populations $i = 1, \ldots, C$, each with probability distribution $N(\mu_i, \Sigma_i)$, where $\theta_i = (\mu_i, \Sigma_i)$ are the unknown mean and covariance of X_{ij}. The likelihood can then be written analogously as

$$P(X|\mu_1, \ldots, \mu_C, \Sigma_1, \ldots, \Sigma_C) = \prod_{i=1}^{C} \prod_{j=1}^{N_i} \Phi(X_{ij}|\mu_i, \Sigma_i),$$

where $\Phi(\mu, \Sigma)$ is the probability density function (pdf) of a normal distribution with mean μ and variance–covariance matrix Σ.

This then corresponds to the maximum likelihood discriminant rules in the frequentist sense (also known as linear discriminant analysis). In the context of microarray experiments, the feature vector X_{ij} denotes the vector of gene expression intensity values for individual j in population i. To complete the hierarchical formulation we need to specify priors on both (μ_i, Σ_i). One natural choice of conjugate priors is a Gaussian prior on μ_i and a corresponding inverse-Wishart prior on Σ_i. Another convenient, noninformative joint prior for (μ_i, Σ_i) is the Jeffreys prior (Jeffreys, 1946). It can be shown that the Jeffreys prior for this particular case corresponds to $P(\mu_1, \ldots, \mu_C, \Sigma_1, \ldots, \Sigma_C) \propto \prod_{i=1}^{C} |\Sigma_i|^{(p+1)/2}$ (the reader might like to try this as an exercise). As already mentioned, inferences using the Jeffreys prior have strong connections to maximum likelihood inference. The posterior predictive density (5.1) for this class of priors can then be shown to be a multivariate t density (Press, 2003),

$$P(Y_{\text{new}}|\bar{X}_i, S_i, \pi_i) \propto \left[1 + \frac{N_i(X_{\text{new}} - \bar{X}_i)'S_i(X_{\text{new}} - \bar{X}_i)}{(N_i + 1)(N - C)}\right]^{-(N-C+1)/2}$$

where $N = \sum_{i=1}^{C} N_i$, $\bar{X}_i = N_i^{-1} \sum_{j=1}^{N_i} X_{ij}$, and $(N_i - 1)S_i = \sum_j (X_{ij} - \bar{X}_i)$ $(X_{ij} - \bar{X}_i)'$.

For simplicity of exposition, consider a binary classification situation with two populations. Assuming $\Sigma_1 = \Sigma_2$, a frequentist rule is to assign X_{new} to class 1 if $P = [1 + (\pi_1/\pi_2)\exp(-L)]^{-1}$ is greater than 1, class 2 is $P < 1$, and at random if $P = 1$, where $L = \log[f_1(\mathbf{X})/f_2(\mathbf{X})]$ is the log-density ratio and π_i is the prior class probability. In the Bayesian context, with Jeffrey's prior, the classification rule can be shown to be $P_B = [1 + (\pi_1/\pi_2)\exp(-L_B)]^{-1}$, where $L_B = \frac{1}{2}\{(\nu + 1)\log[(\nu + r_2\hat{\delta}_2)/(\nu + r_1\hat{\delta}_1)] + p\log(r_1/r_2)\}$, $\hat{\delta}_i = (X - \bar{X}_i)'\hat{\Sigma}^{-1}(X - \bar{X}_i)$, $\hat{\Sigma} = [(N_1 - 1)S_1 + (N_2 - 1)S_2]/(N_1 + N_2 - 2)$, $r_i = N_i/(N_i + 1)$ and $\nu = N_1 + N_2 - 2$. From a fully Bayesian viewpoint, $P_B = P(X_{\text{new}} \in \pi_i|\mathbf{X}) = E(P|\mathbf{X})$, the posterior expected value of P given X (Rigby, 1997). Proof of this is left as an exercise.

Note that until now, we have been working with the full $n \times p$ expression matrix X. This might be unwieldy from both a computational and modeling standpoint. One can incorporate feature selection into this framework in straightforward manner via hierarchical modeling. We sketch out a possible method via the introduction of latent indicator variables, which are a popular tool in Bayesian variable selection (See Chapter 3, Section 3.2.2 on feature selection). Let the indicator variable γ_g be defined such that $\gamma_g = 1$ if gene g is included in the model and $\gamma_g = 0$ if gene g is excluded. Thus the number of genes included in the model is given by $\sum_g \gamma_g = K$, which is assumed less than p via specifying a prior on K to control the number of genes included in the model. Choices for K include any discrete prior such as Poisson, discrete uniform or a geometric distribution. To complete the hierarchical formulation we need to assume a prior on the γ_g. A convenient choice is accorded by a Bernoulli distribution.

5.4   Bayesian Regression Based Approaches to Classification

We now turn our attention to regression approaches to classification which can be treated within a regression framework via generalized linear model (GLM) methodology (McCullagh and Nelder, 1989). The GLMs are a natural generalization of the classical linear models and include as special cases linear regression, analysis of variance, logit and probit models, and multinomial response models.

The GLMs have been used in a variety of applications, including microarrays. In a Bayesian context, classification via GLMs has many potential advantages, especially in the microarray context. Using hierarchical modeling approaches, one can embed the basic GLM into a more complex setting in order to address the scientific question at hand. The model building process is thus elucidated via a series of hierarchical steps as described below.

5.4.1 Bayesian Analysis of Generalized Linear Models

As already mentioned, GLMs are extensions of the linear regression model described in the previous chapter. Specifically, they avoid the selection of a single transformation of the data that must achieve the possibly conflicting goals of normality and linearity imposed by the linear regression model which in some cases is unrealistic ,e.g. for binary or count data. The class of GLMs unifies the approaches needed to analyze such data for which either the assumption of a linear relation between $\mathbf{X}$ (covariates) and $\mathbf{y}$ (responses) or the assumption of normal variation is not appropriate. The name GLM stems from the fact that the dependence of $\mathbf{y}$ on $\mathbf{X}$ is partly linear in the sense that the conditional distribution of $\mathbf{y}$ given $\mathbf{X}$ is defined in terms of a linear combination $\mathbf{X}^T \boldsymbol{\beta}$ as

$$\mathbf{y}|\mathbf{X}, \boldsymbol{\beta} \sim P(\mathbf{y}|\mathbf{X}^T \boldsymbol{\beta}).$$

A GLM is usually specified in two stages:

1. The linear predictor, $\boldsymbol{\eta} = \mathbf{X}^T \boldsymbol{\beta}$.

2. The link function $g(\cdot)$ that relates the linear predictor to the mean of response variable: $\mu(\mathbf{X}) = g^{-1}(\boldsymbol{\eta}) = g^{-1}(\mathbf{X}^T \boldsymbol{\beta})$.

Additionally, a random component characterizing the distribution of the response variable can be specified along with a dispersion parameter. Thus, essentially the mean of the distribution of $\mathbf{y}$, given $\mathbf{X}$, is determined by $\mathbf{X}^T \boldsymbol{\beta}$ as $E(\mathbf{y}|\mathbf{X}) = g^{-1}(\mathbf{X}^T \boldsymbol{\beta})$ which for identifiability reasons is a one-to-one function. The sampling model for the data can then be written as

$$P(\mathbf{y}|\mathbf{X}, \boldsymbol{\beta}) = \prod_{i=1}^{n} P(y_i|\mathbf{X}_i^T \beta).$$

5.4.2 Link Functions

Ordinary linear regression is obviously a special case of the GLM where $g(x) = x$ such that $\mathbf{y}|\mathbf{X}, \beta, \sigma^2 \sim N(\mathbf{X}^T \boldsymbol{\beta}, \sigma^2)$. For continuous data that are all positive, the normal model on the logarithmic scale can be used. Note that the regression coefficients $\boldsymbol{\beta}$ have a distinct interpretation since they are directly related to the response variable, but this interpretation becomes harder for other families due the presence of the link function.

Perhaps the most widely used of the GLMs are those for binary or binomial data which shall be our focus in this chapter. Suppose that $y_i \sim \text{Binomial}(n_i, \mu_i)$ with n_i known. The most common approach is to specify the model in terms of the mean of the proportions y_i/n_i rather than the mean of y_i. The logit transformation of the probability of success, $g(\mu_i) = \log\{\mu_i/(1 - \mu_i)\}$, as the link function leads to the ubiquitous logistic regression model. The sampling model then turns out to be

$$P(\mathbf{y}|\boldsymbol{\beta}) = \prod_{i=1}^{n} \binom{n_i}{\mu_i} \left(\frac{e^{\eta_i}}{1 + e^{\eta_i}} \right)^{y_i} \left(\frac{1}{1 + e^{\eta_i}} \right)^{n_i - y_i}.$$

An alternative link function often used for binary data is the probit link function, where $g(\mu) = \Phi^{-1}(\mu)$, which leads to the sampling model

$$P(\mathbf{y}|\boldsymbol{\beta}) = \prod_{i=1}^{n} \binom{n_i}{\mu_i} \{\Phi(\eta_i)\}^{y_i} \{1 - \Phi(\eta_i)\}^{n_i - y_i}.$$

Although in practice the probit and logit links are very similar, they differ only in the extremes of the tails. Other robust link choices include the complementary log-log link: $g(\mu) = \log(-\log(\mu))$.

5.4.3   GLM using Latent Processes

A key idea, both in computing and interpretation of discrete GLMs, is via latent (unobserved) continuous processes. For example, the probit model for binary data, $P(y_i = 1) = \Phi(X_i^T \beta_i)$, is equivalent to the following model on latent scale z_i where $z_i \sim N(X_i^T \beta_i, 1)$ and

$$y_i = \begin{cases} 1, & \text{if } z_i \geq 0, \\ 0, & \text{if } z_i < 0. \end{cases}$$

There are two advantages of viewing the GLM via latent processes. First, the latent process z_i can be given a useful interpretation in terms of the underlying process that generates the discrete response. Second, the latent parameterization admits a computationally efficient estimation procedure via MCMC techniques, especially the Gibbs sampler. We shall use this idea in many of our examples in the subsequent sections. The latent process framework can also be used for logistic regression using a residual component, as we will demonstrate.

5.4.4   Priors and Computation

The main quantity of interest are the β parameters which quantify the strength of the relationship between the response (classes) and predictors (genes). Estimation and inference in the frequentist context usually proceed via maximum likelihood estimation, which does not have a closed form. In Bayesian inference, prior (informative/noninformative) distributions are elicited for β; however, in most cases conjugate priors do not exist and the posterior distribution is not of closed form. Hence, posterior sampling is usually carried out using adaptive rejection Metropolis sampling (Gilks and Wild, 1992) since in most cases the posterior distribution is log-concave. Another attractive sampling approach that we shall explain here is via the use of latent/auxiliary variables (Albert and Chib, 1993; Holmes and Held, 2006) which reduces the sampling procedure to a series of Gibbs steps which we now explain.

5.4.5 Bayesian Probit Regression using Auxiliary Variables

For simplicity we consider a binary classification problem where the response is usually coded as $Y_i = 1$ for class 1 (diseased tissue) and $Y_i = 0$ (normal tissue) for the other class. Our objective is to use the training data $Y = (Y_1, \ldots, Y_n)$ to estimate $\pi_i(X) = P(Y_i = 1|X_i)$, the probability that a sample is diseased. Assume that Y_i are independent Bernoulli distributed so that $Y_i = 1$ with probability $\pi_i(X)$, independent of other $Y_j, \sim j \neq i$. The gene expression levels of the genes can then be related to the response using a probit regression model,

$$P(Y_i = 1|\beta) = \Phi(X_i'\beta),$$

where X_i is the ith row of the matrix X (vector of gene expression levels of the ith sample), β is the vector of regression coefficients and Φ is the normal cumulative distribution function. Following Albert and Chib (1993), n independent latent variables $Z_1, \ldots, Z_n$ are introduced, with each $Z_i \sim N(X_i'\beta, \sigma^2)$ and where

$$Y_i = \begin{cases} 1, & \text{if } Z_i > 0, \\ 0, & \text{if } Z_i < 0. \end{cases}$$

Hence, we assume the existence of a continuous unobserved or latent variable underlying the observed categorical variable. When the latent variable crosses a threshold (assumed to be 0 here), the observed categorical variable takes on a different value. The key to this data augmentation approach is to transform the model into a normal linear model conditional on the latent responses, as now $Z_i = X_i'\beta + \epsilon_i$ where $\epsilon_i \sim N(0, \sigma^2)$: the residual random effects. The residual random effects account for the unexplained sources of variation in the data, most probably due to explanatory variables/genes not included in the study. We fix $\sigma^2 = 1$ as it is unidentified in the model. It can be seen with this formulation the marginal distribution of Y_i is $\Phi(X_i'\beta)$. The full hierarchical model is now specified as

$$Y_i|Z_i, \beta \sim I(Z_i > 0)\delta_1,$$

$$Z_i \sim N(X_i\beta, 1),$$

$$\beta \sim N(\mu, \sigma^2 V),$$

$$\sigma^2 \sim IG(a, d),$$

where δ_1 is the indicator variable. The prior mean μ on β is usually set to 0, in order to shrink the (effects of) nonimportant genes to 0. V is the prior variance–covariance matrix of the genes, which can modeled in a variety of ways. It can be taken to be diagonal, assuming a priori independence between genes (which is unrealistic in some sense), or can be structured to all correlations/interactions between genes. We will revisit this issue later in the chapter. The MCMC sampling proceeds by sampling all parameters of the model conditional on the Z_i and then sampling Z_i conditional on Y_i from a truncated normal distribution,

$$P(Z_i | Y_i = 1, \beta) \propto P(Y_i = 1 | Z_i, \beta) P(Z_i | \beta)$$

$$= I(Z_i > 0) N(X_i' \beta, 1).$$

Analogously, $P(Z_i | Y_i = 0, \beta) = I(Z_i < 0) N(X_i' \beta, 1)$. Efficient sampling algorithms exist to sample from a truncated normal distribution (Robert, 1995). Conditional on Z_i, all the other model parameters (β, σ^2) have closed-form posteriors (NIG) and can be sampled via the Gibbs sampler.

Future samples can be classified once we have obtained the posterior samples for all the model parameters and random variables $\mathcal{M} = (Z, \beta, \sigma^2)$. Suppose our training data arise (without loss of generality) from the first M samples. Given the sampled parameters from the posterior distribution based on our training data, we sample Z_{M+1}. If the estimated $\widehat{Z}_{M+1} > 0$ then $\widehat{Y}_{M+1} = 1$, and 0 otherwise. The error rates can be estimated via standard methods: misclassification error rates and cross-validation. We delve into this issue in Section 5.6.

In the context of microarrays, as with most classifiers, it is imperative to reduce the dimensionality of the model space before fitting the classifiers. Once can pre-select a certain number of genes based on some criteria (e.g. two-sample t-tests and selecting genes that are above a threshold defined by FDR-based cutoffs) before fitting the above classifier. The disadvantage of this approach, as already mentioned, is that the uncertainty in the selection is not carried forward in the classification mechanism. Bayesian hierarchical machinery gives a coherent way to do this by adding another (gene) selection layer as a part of the general classifier building process. Following Lee et al. (2003) a Gaussian mixture prior for β enables us to perform a variable selection procedure. Define $\gamma = (\gamma_1, \ldots, \gamma_p)$, such that $\gamma_i = 1$ or 0 indicates whether the ith gene is selected or not ($\beta_i = 1$ or 0). Given γ, let $\beta_\gamma = \{\beta_i \in \beta : \beta_i \neq 0\}$ and $\mathbf{X}_\gamma$ be the columns of $\mathbf{X}$ corresponding to β_γ. The prior distribution on β_γ is taken as $N(0, c(\mathbf{X}_\gamma^T \mathbf{X}_\gamma)^{-1})$ with c as a positive scalar (usually between 10 and 100 (see Smith and Kohn, 1996). The indicators γ_i are assumed to be a priori independent with $P(\gamma_i = 1) = \pi_i$, which are chosen to be small to restrict the number of genes. The Bayesian hierarchical model for gene selection can then be summarized as

$$Z|\beta_\gamma, \gamma \sim N(\mathbf{X}_\gamma \beta_\gamma, \sigma^2),$$

$$\beta_\gamma|\gamma \sim N(0, c(\mathbf{X}_\gamma^T \mathbf{X}_\gamma)^{-1}),$$

$$\gamma_i \sim \text{Bernoulli}(\pi_i).$$

The relative importance of each gene can be assessed by the frequency of its appearance in the MCMC samples, i.e. the number of times the corresponding component of γ is 1. This gives us an estimate of the posterior probability of inclusion of a single gene as a measure of the relative importance of the the gene for classification purposes.

5.5   Bayesian Nonlinear Classification

Although conceptually simple and easily implementable, the methods outlined in the previous sections assume a linear relationship between the response and predictor, i.e. $Z_i = X_i'\beta + \epsilon_i$. There are two potential disadvantages of this approach, especially for microarray data: first, they fail to capture nonlinear relationships, if any, present in the data; and second, the genes are assumed independent (additive) in the classification scheme. As with many biological data and especially with gene expression data, this might be a potentially restrictive assumption. We review two methods here that relax each of these assumptions via a nonlinear modeling framework.

5.5.1   Classification using Interactions

We shall use the same notation as above, where Y represents the categorical response (class labels) and $\mathbf{X}$ is the gene expression matrix. Our objective is to use the training data $\mathbf{Y} = (Y_1, \ldots, Y_n)^T$ to estimate $P(\mathbf{X}) = P(\mathbf{Y} = 1|\mathbf{X})$ or alternatively (say) the logit function $f(\mathbf{X}) = \log[P(\mathbf{X})/(1 - P(\mathbf{X}))]$. One could potentially use any link function as mentioned before, but we use the logistic link function to aid interpretability, as we shall see later in the section.

As before, we assume that the Y_i are independent Bernoulli random variables with $P(Y_i = 1) = \pi_i$, so that $P(Y_i|\pi_i) = \pi_i^{Y_i}(1 - \pi_i)^{1-Y_i}$. We construct a hierarchical Bayesian model for classification, again exploiting the latent variable mechanism explained in the previous section. Write $\pi_i = \exp(Z_i)/[1 + \exp(Z_i)]$, where the Z_i are the latent variables introduced in the model to make the Y_i conditionally independent given the Z_i. Z_i is related to $f(\mathbf{X}_i)$ by

$$Z_i = f(\mathbf{X}_i) + \epsilon_i, \tag{5.2}$$

where $\mathbf{X}_i$ is the ith row of the gene expression data matrix $\mathbf{X}$ (vector of gene expression levels of the ith sample) and the ϵ_i are residual random effects. The residual random effects account for the unexplained sources of variation in the data, most probably due to explanatory variables (genes) not included in the study.

The key here is modeling the unknown function $f(\mathbf{X})$ since it is of potentially very high dimension. We would want f to flexible enough to aid accurate prediction and at the same time have a parsimonious structure to aid interpretation. A flexible class of models is accorded via nonparametric regression techniques using basis function expansions,

$$f(\mathbf{X}_i) = \sum_{i=1}^{k} \beta_j B(\mathbf{X}_i, \theta_j),$$

where β are the regression coefficients for the bases $B(\mathbf{X}_i, \theta_j)$, which are nonlinear functions of $\mathbf{X}_i$ and θ. There are various choices of basis functions depending on the scientific question at hand. Examples include regression splines, wavelets, artificial neural networks, and radial bases; see Denison et al. (2002) for a detailed exposition of choices of basis function. An attractive alternative for high-dimensional data is provided by the multivariate adaptive regression spline (MARS) basis function proposed by Friedman (1991). MARS is a popular method for flexible regression modeling of high-dimensional data. Although originally proposed for continuous responses, it has been extended to deal with categorical responses by Kooperberg et al. (1997) and in the Bayesian framework by Holmes and Denison (2003).

The function form of the basis function $f(\mathbf{X})$ is given by

$$f(\mathbf{X}_i) = \beta_0 + \sum_{j=1}^{k} \beta_j \prod_{l=1}^{z_j} (X_{id_{jl}} - \theta_{jl})_{q_{jl}}, \tag{5.3}$$

where k is the number of spline basis, $\beta = \{\beta_1, \ldots, \beta_k\}$ are the set of spline coefficients (or output weights), z_j is the interaction level (or order) of the jth spline, θ_{jl} is a spline knot point, d_{jl} indicates which of the p predictors (genes) enters into the lth interaction of the jth spline, $d_{jl} \sim \, \in \, \sim \{1, \ldots, p\}$, and q_{jl} determines the orientation of the spline components, $q_{jl} \in \{+, -\}$ where $(a)_+ = \max(a, 0)$, and $(a)_- = \min(a, 0)$.

Although apparently complex, the MARS basis function has several advantages in a microarray context:

1. The term z_j determines the order of interaction for a given gene set. Hence genes can enter the MARS basis as a main effect (linear term, $z_j = 1$) or as a bivariate interaction (product term, $z_j = 2$). Higher-order interactions are also possible ($z_j = 3, 4, \ldots$) but are not usually considered due to computational issues and lack of clear interpretability. Thus, depending on k, the number of genes (sets) the MARS model is an additive function of main effect and bivariate interaction terms. This feature models not

only the interactions between the genes but also could could capture the potential nonlinear relationship between the predictor and response.

2. Since the MARS model clearly delineates the relationship between the predictors and the response, this results in much more interpretable models as compared to, say, black box techniques such as artificial neural networks. Such rule-based methods for classification are very useful when determining the effect of individual genes on the response, which in microarrays are usually the odds of having cancer or not.

3. Since the number of splines (genes) is not fixed a priori, variable (gene) selection is built within this framework, thus discerning relevant genes for classification.

4. The MARS model not only captures an interaction between the genes (if any) but also the functional form of the interaction in a flexible manner by positing a spline structure on the product(s) of genes.

Although the MARS model comes with many advantages, MCMC sampling is not trivial: first, the model has a lot of parameters; and second, standard MCMC algorithms such as the Gibbs and Metropolis–Hastings samplers are not applicable here since the model dimension is not fixed: the number of genes/splines entering the model changes at each iteration of the MCMC chain. The first drawback can easily the handled via the latent variable framework, as in (5.2) that reduces most of the conditional distributions, especially of the MARS basis parameters, to a standard form. For the latter one needs a transdimensional algorithm such as the reversible jump MCMC method (Green 1995) for sampling from the posterior distribution.

The priors on the model parameters are as follows. A Gaussian prior is assumed for β with mean 0 and variance Σ. The form of Σ is usually taken to be diagonal to aid parsimony in an already overparameterized model. Specifically $\Sigma = \sigma^2 \mathbf{D}^{-1}$, where $\mathbf{D} \equiv \mathrm{diag}(\lambda_1, \lambda, \ldots, \lambda)$ is an $(n+1) \times (n+1)$ diagonal matrix. We fix λ_1 to a small value, amounting to a large variance for the intercept term, but keep λ unknown. We assign a inverse-gamma prior to σ^2 and a gamma prior to λ with parameters (γ_1, γ_2) and (τ_1, τ_2), respectively. The prior structure on the MARS model parameters is as follows. The prior on the individual knot selections θ_{jl} is taken to be uniform over the n data points $P(\theta_{jl}|d_{jl}) = U(x_{1d_{jl}}, x_{2d_{jl}}, \ldots, x_{nd_{jl}})$, where d_{jl} indicates which of the genes enter our model and $p(d_{jl})$ is uniform over the p genes, $P(d_{jl}) = U(1, \ldots, p)$. The prior on the orientation of the spline is again uniform, $P(q_{jl} = +) = P(q_{jl} = -) = 0.5$. The interaction level in each spline has a prior, $P(z_j) = U(1, \ldots, z_{max})$, where z_{max} is the maximum level of interaction set by the user. Finally, the prior on k, the number of splines, is taken to an improper one, $P(k) = U(1, \ldots, \infty)$, which indicates no a priori knowledge on the number of splines. Hence, the model now has only one user defined parameter, z_{max}, the maximum level of interaction, which is usually set to 2, as mentioned earlier in this section.

As already suggested, conventional MCMC methods such as the Metropolis–Hastings algorithm (Metropolis et al., 1953; Hastings, 1970) are not applicable here since the parameter (model) space is variable: we do not know the number of splines (genes) a priori. Hence, we use the variable dimension reversible jump algorithm outlined in Green (1995) and modified for our framework by Denison et al. (2002). Specifically the chain is updated using the following proposals with equal probability: (1) add a new spline basis to the model; (2) remove one of the k existing spline bases from the model; (3) alter an existing spline basis in the model (by changing the knot points). Conditional on k, efficient sampling can then be achieved by the use of latent variables. For more details of the procedure, see Baladandayuthapani et al. (2006).

5.5.2   Classification using Kernel Methods

Another closely related nonlinear classifier is based on the Bayesian reproducing kernel Hilbert spaces (RKHS) approach (Aronszajn, 1950; Parzen, 1970) proposed by Mallick et al. (2005). RKHS are especially useful for microarray data since they project the feature space into a space of dimension much less than p, the number of genes. Classical RKHS use decision-theoretic framework with no explicit probability model, hence it is not possible to assess the uncertainty of both the classifier and the predictions based on it. Mallick et al. (2005) propose a full probabilistic model-based approach to RKHS classification which has several advantages, as explained below.

Consider the binary classification problem with two classes. The data construct contains $(Y, \mathbf{X})$, where $Y \in \{0, 1\}$ codes the class labels from n training samples and $\mathbf{X}$ is the $n \times p$ matrix of gene expression values. Classification proceeds by specifying a probability model on $P(Y|\mathbf{X})$. As before, introduce latent variables $Z = (Z_1, \ldots, Z_n)$, conditional on which the $P(Y|Z) = \prod_{i=1}^{n} P(Y_i|Z_i)$, i.e. the Y_i are conditionally independent given the Z_i. The latent variables are then linked to covariates $\mathbf{X}$ by $Z_i = f(X_i) + \epsilon_i$ for $i = 1, \ldots, n$, where X_i is the gene expression vector associated with the ith array. To complete the model specification one needs to specify $P(Y|Z)$ and f.

To model $P(Y|Z)$, one can exploit the duality between *loss* and *likelihood*, whereby the loss can be viewed as the negative of log-likelihood. If the loss function is defined as $l(y, Z)$, then minimizing this loss function corresponds to maximizing $\exp\{-l(y, z)\}$, which is proportional to the likelihood function. There are various choices of loss and corresponding likelihood functions available.

As in the Bayesian MARS (BMARS) classifier, the key here again is modeling the high-dimensional function $f(\mathbf{X})$, to keep it flexible and at the same time admit parsimony. This can achieved by modeling f via an RKHS approach. A Hilbert space $\mathcal{H}$ is a collection of functions on a set T with an associated inner product $\langle g_1, g_2 \rangle$ and the norm $\|g_1\| = \langle g_1, g_2 \rangle$ for $g_1.g_2 \in \mathcal{H}$. An RKHS $\mathcal{H}$ with reproducing kernel K is a Hilbert space having an associated function K on $T \times T$ with the properties

(a) $K(\cdot, X) \in \mathcal{H}$,

(b) $\langle K(\cdot, X), g(\cdot) \rangle = g(X)$ for all $X \in T$ and for every g in $\mathcal{H}$,

where $K(\cdot, X)$ and $g(\cdot)$ are functions that are defined on T with values $X^* \in T$ equal to $K(X^*, X)$ and $g(X^*)$, respectively. The reproducing kernel function provides a basic building block for $\mathcal{H}$ via the following lemma (shown here without proof) from Parzen (1970).

Lemma 5.1. If K is a reproducing kernel for the Hilbert space $\mathcal{H}$, then $\{K(, \cdot, X)\}$ spans $\mathcal{H}$.

There are two important consequences of this lemma. First, any $g \in \mathcal{H}$ with reproducing kernel K can approximated to any desired level of accuracy via a dense function in $\mathcal{H}$ given by $g_N(\cdot) = \sum_{j=1}^{N} \beta_j K(\cdot, x)j)$, where $x_j \in T$ for each $j = 1. \ldots, N$. Second and more importantly, for gene expression data, we achieve dimension reduction as follows. Given the expression vectors $X_1, \ldots, X_n$ and Z the latent variables, let $f \in \mathcal{H}$ with choices of K discussed below. The optimal f is based on minimizing $\sum_{i=1}^{n} \{z_i - f(X_i)\}^2 + \|f\|^2$ with respect to f which must admit the representation (Wahba, 1990)

$$f(\cdot) = \beta_0 + \sum_{j=1}^{n} \beta_j K(\cdot, X_j). \tag{5.4}$$

This reduces the dimensionality of the optimization to n, which is far less than p, the number of genes. Inference on f is made via inference on $\beta = (\beta_0, \ldots, \beta_n)^T$. The choice of kernels can be flexible by allowing them to depend on unknown parameters. Classical choices include the Gaussian kernel $K(X_i, X_j) = \exp\{-\|X_i - X_j\|^2/\theta\}$ and the polynomial kernel $K(X_i, X_j) = \{(X_i.X_j) + 1\}^\theta$, where $X.Y$ denotes the inner product of two vectors X and Y.

We now show how we can exploit the above framework to achieve classification via Bayesian suport vector machines (SVMs). We refer the reader to Cristianini and Shawe-Taylor (2000), Schölkopf and Smola (2002) and Herbrich (2002) for the basics of SVMs. To follow notation, suppose the class labels are coded as $Y_i = 1$ and $Y_i = -1$ for the two classes. The basic idea behind SVMs is to find a linear hyperplane that best separates the observations with $Y_i = 1$ from those with $Y_i = 1$ called the *margin*. As shown by Wahba (1999) or Pontil et al. (2000), this optimization problem amounts to finding β so as to minimize $\|\beta\|^2 + C \sum_{i=1}^{n} [1 - Y_i f(X_i)]_+$ where $[a]_+ = a$ if $a > 0$ and $[a]_+ = 0$ otherwise. $C \geq 0$ is a penalty term and f is given by (5.4). The problem can be solved by using nonlinear programming methods. Fast algorithms for computing SVM classifiers can be found in Chapter 7 of Cristianini and Shawe-Taylor (2000).

In the Bayesian paradigm, this optimization problem is equivalent to finding the posterior mode of β, where the likelihood is given by $\exp[-\sum_{i=1}^{n}[1 - Y_i f(X_i)]_+$ with a $N(\mathbf{0}, C I_{n+1})$ prior on β where we exploit the loss–likelihood duality mentioned above. With the formulation based on latent variables Z, the density can be written as

$$P(Y|Z) \propto \exp\left(-\sum_{i=1}^{n}[1 - Y_i Z_i]_+\right), \qquad (5.5)$$

with the prior on $Z_i \sim N\{f(X_i), \sigma^2\}$. Writw (5.4) in matrix notation as $f(X_i) = K'_i\beta$, where K is a positive definite function of the covariates/genes X and some unknown parameter θ. A Gaussian prior is assumed over β with mean $\mathbf{0}$ and variance $\sigma^2 D$, where $D \equiv \mathrm{diag}(\lambda_1, \lambda, \ldots, \lambda)$ is of size $n \times n$. The precision λ_1 of the intercept term is set to small number, corresponding to a large variance. A proper uniform prior is assumed over θ, a conjugate inverse-gamma prior over σ^2 and λ^{-1}.

Note, however, that the density in (5.5) may involve Z. Following Sollich (2001), one may bypass this problem by assuming a distribution for Z such that the normalizing constant cancels out. If the normalized likelihood is is written as $\exp\left(-\sum_{i=1}^{n}[1 - Y_i Z_i]_+\right)/c(Z)$, where $c(\cdot)$ is the normalizing constant, then, choosing $P(Z) \propto Q(Z)c(Z)$, the joint distribution turns out to be

$$P(Y|Z) \propto \exp\left(-\sum_{i=1}^{n}[1 - Y_i Z_i]_+\right) Q(Z),$$

as $c(\cdot)$ cancels from the expression. A simple choice of $Q(Z)$ is a product of independent normal probability density functions with means $f(X_i)$ and common variance σ^2. Mallick et al. (2005) refer to this method as Bayesian support vector machine (BSVM) classification. We refer the reader to this article for further details.

5.6   Prediction and Model Choice

The main goal of classification is the prediction of new samples. In the Bayesian context this can be achieved via the posterior predictive densities. For a new sample with gene expression matrix $\mathbf{X}_{\mathrm{new}}$, the marginal posterior distribution of the new sample Y_{new} is given by

$$P(Y_{\mathrm{new}} = 1|\mathbf{X}_{\mathrm{new}}) = \int_{\mathcal{M}} P(Y_{\mathrm{new}} = 1|\mathbf{X}_{\mathrm{new}}, \mathcal{M}, Y)P(\mathcal{M}|Y)d\mathcal{M}, \qquad (5.6)$$

where $\mathcal{M}$ contains all the unknowns in the model, $P(\mathcal{M}|Y)$ is the posterior probability of the unknown random variables and parameters in the model. Assuming conditional independence of the responses, the integral in (5.6) reduces to

$$\int_{\mathcal{M}} P(Y_{\mathrm{new}} = 1|\mathbf{X}_{\mathrm{new}}, \mathcal{M})P(\mathcal{M}|Y)d\mathcal{M}, \qquad (5.7)$$

with the associated measure uncertainty being $P(Y_{new} = 1|\mathbf{X}_{new}, Y)\{1 - P(Y_{new} = 1|\mathbf{X}_{new}, Y)\}$. The integral given in (5.8) is computationally and analytically intractable and needs approximation procedures. We approximate it by its Monte Carlo estimate,

$$P(Y_{new} = 1|\mathbf{X}_{new}) = \frac{1}{m} \sum_{j=1}^{m} P(Y_{new} = 1|\mathbf{X}_{new}, \mathcal{M}^{(j)}), \qquad (5.8)$$

where $\mathcal{M}^{(j)} \sim$ for $\sim j = 1, \ldots, m$ are the m MCMC posterior samples of the model unknowns $\mathcal{M}$. The approximation (5.8) converges to the true value (5.7) as $m \to \infty$.

In order to select from different models/classifiers, we generally use misclassification error. When a test set is provided, we first obtain the posterior distribution of the parameters based on training data, Y_{trn} (train the model), and use them to classify the test samples. For a new observation from the test set, $Y_{i,test}$ we will obtain the probability $P(Y_{i,test} = 1|Y_{trn}, X_{i,test})$ by using the approximation to (5.8) given by (5.7). If there is no test set available, we will use a hold-one-out cross-validation approach. We follow the technique described in Gelfand (1996) to simplify the computation. For the cross-validation predictive density, in general, let $\mathbf{Y}_{-i}$ be the vector of Y_js without the ith observation Y_i,

$$P(Y_i|\mathbf{Y}_{-i}) = \frac{P(\mathbf{Y})}{P(\mathbf{Y}_{-i})} = \left[\int \{P(y_i|\mathbf{Y}_{-i}, \mathcal{M})\}^{-1} P(\mathcal{M}|\mathbf{Y}) d\mathcal{M} \right]^{-1}.$$

The MCMC approximation to this is

$$\widehat{P}(Y_i|\mathbf{Y}_{-i,trn}) = m^{-1} \sum_{j=1}^{m} \left\{ P(y_i|\mathbf{Y}_{-i,trn}, \mathcal{M}^{(j)}) \right\}^{-1},$$

where $\mathcal{M}^{(j)} \sim$ for $\sim j = 1, \ldots, m$ are the m MCMC posterior samples of the model unknowns $\mathcal{M}$. This simple expression is due to the fact that the Y_i are conditionally independent given the model parameters $\mathcal{M}$. If we wish to make draws from $p(Y_i|\mathbf{Y}_{i,trn})$, then we need to use importance sampling.

5.7 Examples

We illustrate some of the methods discussed above with some real experimental data. We use the microarray data set used in Hedenfalk et al. (2001), on breast tumors from patients carrying mutations in the predisposing genes, BRCA1 or BRCA2, or from patients not expected to carry a hereditary predisposing mutation. Pathological and genetic differences appear to imply different but overlapping functions for BRCA1 and BRCA2. They examined 22 breast tumor samples from 21 breast cancer patients, and all patients except one were women. Fifteen women had hereditary breast cancer, 7 tumors with BRCA1 and

8 tumors with BRCA2. For each breast tumor sample 3226 genes were used. We use our method to classify BRCA1 versus the others (BRCA2 and sporadic). As a test data set is not available, we have used a full leave-one-out cross-validation test and use the number of misclassifications to compare the various approaches. Table 5.1 summarizes our results. BLDA stands to Bayesian linear discriminant analysis, and BPR for Bayesian probit regression.

Table 5.1   Model misclassification errors using hold-one-out cross-validation for breast cancer data

Model	Number of misclassifies samples
BLDA	1
BPR	0
BMARS	0
BSVM	0
Probabilistic neural network ($r = 0.01$)	3
k nearest neighbors ($k = 1$)	4
SVM linear	4
Perceptron	5

We compare the cross-validation results with other popular classification algorithms including feed-forward neural networks (Williams and Barber, 1998), k nearest neighbours (Fix and Hodges, 1951), classical SVMs (Vapnik, 2000), perceptrons (Rosenblatt, 1962) and probabilistic neural networks (Specht, 1990) in Table 5.1 All these methods have used expression values of only 51 genes as used in Hedenfalk et al. (2001), thus pre-selecting the genes in a two-step approach. Note that the Bayesian methods perform better than the other methods, at least for this example. Moreover, the regression-based approaches achieve almost perfect classification for this data set, while BLDA misclassifies one sample. When we used BLDA without a variable selection procedure it performed poorly, misclassifying four samples, thus emphasizing the importance of feature selection before classification for microarray data. As an example we present a list of top 25 genes that were selected using the BMARS model in Table 5.2. The list is sorted by the number of times the gene appears in the MARS basis function in the MCMC sampler – which gives some indication of the relative importance of each gene. The asterisk (*) corresponds to the common genes that are selected using a BPR. Another nice feature of Bayesian approaches is that as a byproduct we also get an (honest) estimate of the uncertainty in classification. For example, with BSVM procedure the 95% credible interval for misclassification was (0, 3).

Table 5.2   Breast cancer data: top 25 genes from the BMARS model. The asterisk(*) corresponds to the common genes that appear by using a linear probit classifier

Image Clone ID	Gene description
767817	polymerase (RNA) II (DNA directed) polypeptide F
307843	ESTs (*)
81331	'FATTY ACID-BINDING PROTEIN, EPIDERMAL'
843076	signal transducing adaptor molecule (SH3 domain and ITAM motif) 1
825478	zinc finger protein 146
28012	O-linked N-acetylglucosamine (GlcNAc) transferase
812227	'solute carrier family 9 (sodium/hydrogen exchanger), isoform 1'
566887	heterochromatin-like protein 1 (*)
841617	ornithine decarboxylase antizyme 1 (*)
788721	KIAA0090 protein
811930	KIAA0020 gene product
32790	'mutS (E. coli) homolog 2 (colon cancer, nonpolyposis type 1)'
784830	D123 gene product (*)
949932	nuclease sensitive element binding protein 1 (*)
26184	'phosphofructokinase, platelet' (*)
810899	CDC28 protein kinase 1
46019	minichromosome maintenance deficient (S. cerevisiae) 7 (*)
897781	keratin 8 (*)
32231	KIAA0246 protein (*)
293104	phytanoyl-CoA hydroxylase (Refsum disease) (*)
180298	protein tyrosine kinase 2 beta
47884	macrophage migration inhibitory factor (glycosylation-inhibiting factor)
137638	ESTs (*)
246749	'ESTs, weakly similar to trg [R. norvegicus]'

5.8   Discussion

Since the explosion of microarrays in the past decade there have been many approaches to classification using gene expression profiling. We have reviewed some of recent developments in Bayesian classification of gene expression data, both from a linear and a nonlinear perspective, with special emphasis on regression-based methods for classification. We also hasten to point out that there is a growing literature on classification methods from both a Bayesian and frequentist standpoint for microarray data. Some additional methods not covered in this book include the classification methods based on Bayesian tree models (Ji et al., 2006), factor regression models (West, 2003), and Bayesian nonparametric approaches based on Dirichlet process priors.

We have treated the binary classification in detail. When the response is not binary, such that the number of classes (C) is greater than two, then the problem becomes a multiclass classification problem. This can be handled in a manner similar to the binary classification approach, as follows. Let $\mathbf{Y}_i = (Y_{i1}, \ldots, Y_{iC})$ denote the multinomial indicator vector with elements $Y_{iq} = 1$ if the qth sample belongs to the qth class and $Y_{ij} = 0$ otherwise. Let $\mathbf{Y}$ denote the $n \times C$ matrix of these indicators. The likelihood of the data given the model parameters $(\Theta_1, \ldots, \Theta_C)$ is given by

$$P(Y_i = 1 | \mathbf{X}_i) = \pi_1^{y_{i1}} \pi_2^{y_{i2}} \ldots \pi_C^{y_{iC}},$$

where π_q is the probability that the sample came from class q. This is modeled in a similar manner to the binary class case as discussed above.

6

Bayesian Hypothesis Inference for Gene Classes

6.1   Interpreting Microarray Results

In Chapter 3, we discussed the many challenges of detecting differentially expressed genes. The Bayesian approach provides very powerful and often superior methods for inferring differential expression. Gene expression experiments generate an overwhelming quantity of expression measurements on typically thousands of genes. Producing a interesting list of genes associated with a phenotype or outcome is an important accomplishment, but by no means final. The results must be interpreted in such a way that biological conclusions can be drawn from the experiment. Here we adopt the strategy of integrating biological knowledge and experimental design in a powerful but flexible way.

Making sense of the biological roles of genes from large gene expression experiments requires insight into the biological regulation of genes at the molecular level. Drawing biological conclusions from gene expression arrays requires more than a cursory understanding of the relevant biology. A familiarity with cell biology, organic chemistry, and biochemistry is helpful. Suppose that a particular expression pattern is observed in a series of bio-markers in a known relevant pathway. Genes showing similar experimental profiles, it has been hypothesized, perform related functions in the cell. In practice this hypothesis may or may not be true, on a case by case basis. In other words, genes can show high association in expression, and be physically unrelated, or so dissociated through a chain of molecular events as to negate any meaningful biological associations. The converse can also be true. Gene expression can be highly physically dependent, but

Bayesian Analysis of Gene Expression Data   B. Mallick, D. Gold, and V. Baladandayuthapani
© 2009 John Wiley & Sons, Ltd

in ways so nontrivial that it is unlikely that a naive co-expression analysis will infer an association. While these facts present difficulties, there are nevertheless benefits to making use of historical prior information.

Commonly available information of regulatory relationships or shared pathways, i.e. the larger biological context, is used in gene expression analyses in order to interpret results in a biologically meaningful way. Gene–gene co-expression, for example, for genes known to be involved in, say, photosynthesis or metabolite regulation, is conceptually helpful, more so than, say, the collection of a few hundred or even thousand gene-wise events, where little is known about the genes. This implies that working with, albeit limited, biological information can shift the emphasis of an analysis from a collection of gene-wise variables to a class of documented biological events, associated with a particular cellular task or function, potentially leading to discovery of an important class of epistatic events.

The aim of gene class detection is to perform inference on a collections of genes, defined by an historical source. Experimental design and biological knowledge are the keys to drawing biologically meaningful and logical conclusions. We must condition any conclusions on the given biology, as our understanding is incomplete. In this chapter we adopt the strategy of integrating biological knowledge and experimental design in a powerful but flexible way.

6.2   Gene Classes

Historical information concerning genes and their related biological roles is readily available in the form of on-line databases that can be mined with sophisticated queries. Much of this information is publicly accessible, following academic and government initiatives to promote accessibility as well as high standards for analysis of high-throughput biology. There are many ways to utilize this information. In this chapter we consider using such information by defining collections of genes, or biological clusters of genes, based on biological characteristics. A *gene class* is here defined as a collection of genes defined to be biologically associated given a biological reference such as gene ontology, historical pathway databases, scientific literature, transcription factor databases, expert opinion, empirical evidence (e.g. past experiments), and theoretical evidence.

The Gene Ontology (GO) Consortium offers one of the most detailed, complete, and up-to-date sets of gene-annotations available. It consists of 17 members, each responsible for annotating and curating the gene annotations of a class of model organisms, ranging from protozoa to *Homo sapiens*. The GO database is publicly available at http://www.geneontology.org. The GO annotations consist of a dense hierarchically structured vocabulary describing three components of gene products: the biological processes (what), molecular functions (how), and cellular components (where). The very hierarchical structure of each of the three components of the language can be visualized as a separate directed acyclic graph (DAG). The nodes along a single branch of a graph provide multiple levels of description of each gene products. For example, on the biological process DAG, the child nodes under *transcription* include *positive regulation of transcription* and *negative regulation of transcription*. The annotations are completed by a mapping of the known genes to membership on the nodes of the GO DAGs.

In order to facilitate visualization of GO, there are many publicly available browsers, such as the Cancer Genome Anatomy Project (CGAP) browser at http://cgap.nci.nih.gov/Genes/GOBrowser. A user can explore the dense vocabulary rooted within the three GO directed acyclic graphs. In practice, analysis of GO gene classes is performed on pruned versions of the GO hierarchical DAGs. The vocabulary of the GO categories provides finer detail on the genes, as one moves down the DAG, while the density of the nodes, i.e. the number of genes mapped to them, becomes sparser. The highly resolved level of detail, combined with the sparsity of genes, near terminal nodes, makes their inclusion typically impractical for analysis. New terminal nodes can be derived by determining the maximum depth for each branch along the DAG, annotating gene mapping to nodes beneath the maximum depth to the newly defined terminal leaf nodes. For example, consider *proliferation*. If one wishes to define proliferation as a terminal node, genes mapped to child nodes of proliferation are assigned to the proliferation parent class.

Historical pathway databases largely portray the relationships of genes, transcripts and gene products in the form of diagrams and static graphs. The sources for historical pathways vary from the literature to panels of biological experts. In contrast to GO, the pathway databases tend to have much lower coverage of the genome, with much more detailed information about the genes and how they interact. There are many different kinds of pathways and, as such, many ways in which genes can be associated in a myriad of processes. Genes in a pathway can perform very different roles triggered at very different times. Genes, for example, can be involved in a metabolic pathway, a signaling pathway, or a cascade response to external stimuli in the cell. The nature of the way genes can interact, e.g. direct protein phosphorilation, or DNA binding, makes using such information complicated, as gene expression arrays measure relative transcription levels. It would appear reasonable to build gene classes from transcriptional regulation as gene regulatory factors can explain variation on arrays (Tamada et al., 2003; Allocco et al., 2004; Nagaraj et al., 2004; Bhardwaj and Lu, 2005).

Working with gene classes is intuitively a supervised dimension reduction step, reducing thousands of gene-wise variables to a more sophisticated and informed source of covariates, reflecting historical pathways, biological processes, or regulatory roles. This reduction might be quite useful, although in some cases it can be quite awkward. This information may or may not be useful for microarray analysis, as pathways and processes can be dynamic processes, while microarrays provide a one-time snapshot of relative gene expression. Incorporating extra network structure in an analysis is a nontrivial tasks requiring sensible assumptions. Despite the enormity of past data, and the prior information on gene classes and historical pathways, there are limitations to using this information. While complex, the annotations are incomplete, the implication being that any conclusions must be conditioned upon the current state of progress in annotating the genome. In some organisms, such as yeast and *E. coli*, the coverage of the genomes is high, while in others, such as *H. sapiens*, the coverage is far poorer.

6.2.1   Enrichment Analysis

We begin our discussion of gene class detection, taking as given that gene classes are well defined. Membership in a gene class does not imply co-regulation, or necessarily suggest likely co-expression. We understand that some genes within a gene class can show differential expression, conditional upon the experimental factors, while others do not. What makes gene classes useful is the degree of evidence provided among the genes assigned to the class, that a particular regulatory process or function is activated/deactivated across experimental factors. Each gene provides a vote of evidence in favor of or not in favor of the hypothesis that the process or pathway represented by the gene class is biologically somehow affected by or part of a response to the experimental stimulus or conditions.

Genes can be members of more than one gene class, and therefore statistical measures of differential expression among gene classes can be dependent. Gene class statistics can be dependent for other reasons, such as biology. Another very important consideration is that gene class membership is incomplete. The conclusions of enrichment analysis must be conditioned upon on the annotations, lacking further knowledge.

Suppose a gene expression microarray experiment is conducted to compare gene expression in two groups, experimental and control. Let b be a known class of genes, such as a specific GO biological function. Of the the G unique genes on the array, some proportion show significant differential expression. Table 6.1 shows the counts of genes that are members of gene class b by differential expression.

Table 6.1   Enrichment analysis table, gene class b

	Differentially expressed	Not differentially expressed	Total
Member of b	G_{11}	G_{12}	$G_{1.}$
Not a member of b	G_{21}	G_{22}	$G_{2.}$
Total	$G_{.1}$	$G_{.2}$	G

Enrichment analysis seeks to test the hypotheses

$H_0^{(b)}$: The differentially expressed genes were sampled i.i.d. uniformly at random without replacement from the G genes on the array,

$H_1^{(b)}$: The differentially expressed genes were sampled from gene class b in greater proportion than G_{11}/G,

i.e. the the set of differentially expressed genes is 'enriched' with genes from class b (Mootha et al., 2003). Following the assumption under H_0 that the differentially expressed genes were sampled i.i.d. uniformly at random without replacement from the G genes on the array, the probability of observing G_{11} genes from class b changing is given by the hypergeometric distribution. The p-value for testing H_0 is calculated by the probability of observing a count greater than or equal to G_{11}. This is the procedure that R. A. Fisher (1922) proposed for testing one-way independence of counts in 2×2 contingency tables, the one-way Fisher exact test.

There are many programs available that perform enrichment analysis, among them GO Miner and GOsurfer (for a review, see Curtis et al., 2005). These applications are equipped with sophisticated graphical user interfaces, allowing one to view the results of an enrichment analysis displayed on directly on a GO DAG, with color codes to indicate up or down regulation.

Gene set enrichment analysis (GSEA; Subramanian et al., 2005), extends the ideas of enrichment analysis, incorporating information about the empirical distribution of the gene-wise test statistics. Suppose that an experiment is conducted to compare expression between two groups. In GSEA, test statistics such as t-statistics are computed for each gene. The distribution of t-statistics of genes in each gene class are compared to the distribution of t-statistics in the gene class compliment, by what is essentially a Kolmogorov–Smirnov score statistic. A measure of significance is derived by permuting the subject (array) labels in each experimental group. Gene-wise randomizations have also been suggested by Kim and Volsky (2005), leading to a parametric computational shortcut, although they can perturb gene–gene correlations. Efron and Tibshirani (2006) proposed a more powerful alternative to the Kolmogorov–Smirnov score statistic for inferring gene set enrichment, the maxmean statistic, which is computed as the larger of the absolute value of the average of positive t-statistics or negative t-statistics, for genes in a gene class. As for GSEA, Efron and Tibshirani use a permutation-based approach to derive the null. In order to appropriately account for variation in the null, both the gene class maxmean statistics, and permuted gene class statistics, are normalized, subtracting the respective means and dividing the respective standard deviations, before comparison to derive p-values, a procedure Efron and Tibshirani call restandardization. Another recent Bayesian alternative is provided by Newton et al. (2007).

A multivariate generalization of enrichment analysis (MEA) was proposed by Klebanov (2007). Modifying the enrichment hypotheses extensively, Klebanov considers the correlation structure of genes, comparing the joint distribution of gene profiles between phenotypes, or experimental factors, for genes in each a gene class with what is called the N-statistic. The test is performed, given a set of genes in a gene class, by comparing the joint multivariate distance of gene profiles between phenotype groups with the average multivariate distance of the gene profiles within each phenotype, in each gene. The N-statistic null distribution is derived by permuting subject (array) labels. Unlike enrichment analysis, GSEA and MEA are not performed conditionally upon a set of discovered candidate genes.

6.3   Bayesian Enrichment Analysis

Historical pathways help facilitate inference on collections of genes that may act in concert. Utilizing such information can be a complex task for which Bayesian formalism helps, making prior beliefs explicit as well as challenging them. Suppose that one wishes to draw biological evidence of an association between an entire pathway of genes and an experimental outcome of interest. Inference is drawn from synthesizing what we know about the genes involved in the particular regulatory process or function, i.e. the pathway, with data, observed or experimental.

In a Bayesian enrichment analysis, the parameters indicating differential expression are accounted for by posterior uncertainty. Differential gene expression is inferred with posterior uncertainty, contributing a weight of probability in favor of enrichment, rather than casting a single vote. Let the set $\mathcal{G}_D$ be the collection of genes selected for change, $\mathcal{G}_b$ be the collection of genes with function b, and $||\mathcal{G}||$ be the cardinality of the set $\mathcal{G}$. The variable

$$\Upsilon_b = I(||\mathcal{G}_D \cup \mathcal{G}_b||/||\mathcal{G}_D|| \geq \alpha) \tag{6.1}$$

measures the event that the proportion of differentially expressed genes, $\mathcal{G}_D$, enriched for gene class b, $\mathcal{G}_b$, exceeds a user-specified $\alpha \in (0, 1)$ for indicator function I(A), where I(A) = 1 if the event A is true and 0 otherwise. In Bayesian enrichment analysis, we summarize the posterior distribution of, and make probability statements about, enrichment through Υ_b, e.g. $E(\Upsilon_b|\mathbf{X})$.

Consider defining Υ_b as

$$\Upsilon_b = I(||\mathcal{G}_D \cup \mathcal{G}_b||/||D|| \geq ||\mathcal{G}_D^c \cup \mathcal{G}_b||/||\mathcal{G}_D^c||), \tag{6.2}$$

to indicate that the proportion of genes with function b that are selected exceeds the proportion not selected. This is the definition adopted by Bhattachajee et al. (2001), in a study to infer differential gene expression between organ tissues. One could alternatively define

$$\Upsilon_b = I(||\mathcal{G}_D \cup \mathcal{G}_b||/||\mathcal{G}_b|| \geq ||\mathcal{G}_D||/(||\mathcal{G}_D|| + ||\mathcal{G}_D^c||)). \tag{6.3}$$

These definitions of Υ obviously address different hypotheses, with (6.2) most closely resembling the test of enrichment. Accounting for the uncertainty in gene detection and enrichment of gene classes in probabilistic fashion is quite natural in a Bayesian setting.

Example 6.1 SCC STUDY
We return to the Wachi et al. (2005) data set consisting of 22 283 expression probe values from ten patients, five with SCC (group 1) and five normal lung patients (group 2). The likelihood for gene g is defined for the difference in sample averages $\bar{x}_{g1}$ and $\bar{x}_{g2}$ for the respective groups, and the appropriate standard deviation estimate $\hat{\sigma}_g$, as

$$\frac{\bar{x}_{g1} - \bar{x}_{g2} - \delta_g}{\hat{\sigma}_g \sqrt{(1/5 + 1/5)}} | \delta_g \sim t_8, \tag{6.4}$$

a Student's t-distribution with 8 degrees of freedom, given the unknown mean log-ratio δ_g. Suppose that an improper mixture prior is specified for δ_g

$$P(\delta_g|\lambda_g) \propto \lambda_g 1_{\{\delta_g < -1 \cup \delta_g > 1\}} + (1 - \lambda_g) 1_{\{-1 \le \delta_g \le 1\}},$$

$$P(\lambda_g) = \text{Bernoulli}\left(\frac{1}{2}\right), \tag{6.5}$$

conditional upon the discrete variable $\lambda_g = 0$ indicating the event $\delta_g \in (-1, 1)$, i.e. less than twofold differential expression on the original scale, and $\lambda_g = 1$ indicating that $\delta_g \in (-\infty, -1) \cup (1, \infty)$, or greater than twofold differential expression.

Integrating over δ_g, the marginal posterior of $\lambda_g = 0$ is Bernoulli, with probability

$$\pi_g|\mathbf{x_g} = \int_{-1}^{1} t_8(s; \bar{x}_{g1} - \bar{x}_{g2}, \hat{\sigma}_g^2 (1/5 + 1/5))ds, \tag{6.6}$$

the integral of a t random variable from -1 to 1, with location $\bar{x}_{g1} - \bar{x}_{g2}$ and dispersion $\hat{\sigma}_g^2 (1/5 + 1/5)$. Defining Υ_b as in (6.2), we obtain a the posterior probability of enrichment, sampling λ_g for $g = 1, \ldots, G$ and at each iteration, and deriving counts $||\mathcal{G}_b||$, $||\mathcal{G}_D||$, $||\mathcal{G}_D^c||$ and $||\mathcal{G}_b^c||$, and subsequently Υ_b. The gene classes were obtained from GO biological process annotations. The top 25 gene classes, ranked by the posterior expectation of Υ_b are listed in Table 6.2.

6.4 Multivariate Gene Class Detection

Multivariate gene class detection seeks to infer differential expression among genes and gene classes. The collection of methods available for multivariate gene detection and gene class inference is quite diverse. Some methods incorporate gene class information in the modeling step, others derive measures of enrichment from modeling results. The proposed methods can also vary to the degree to

Table 6.2   Bayesian enrichment analysis, Wachi data

Gene class	No. of genes	Posterior expectation Υ_b (6.2)
chloride channel activity	68	1
cytokine activity	126	1
endopeptidase inhibitor activity	122	1
extracellular matrix structural constituent	140	1
GPI anchor binding	147	1
heme binding	161	1
heparin binding	129	1
integrin binding	85	1
NAD binding	56	1
oxygen binding	59	1
protein binding, bridging	52	1
serine-type endopeptidase inhibitor activity	89	1
structural constituent of cytoskeleton	155	1
transmembrane receptor protein tyrosine kinase activity	51	1
chemokine activity	53	0.9999
electron carrier activity	132	0.9999
sugar binding	178	0.9994
G-protein coupled receptor activity	88	0.9992
growth factor binding	65	0.9989
actin filament binding	79	0.9961
protein kinase inhibitor activity	50	0.994
protein tyrosine phosphatase activity	146	0.9911
kinase inhibitor activity	52	0.9792
ATPase activity, coupled to transmembrane movement of substances	51	0.9764
lipid transporter activity	68	0.9764

which dependence is assumed between the genes in a gene class, and between gene classes, inferring differential expression among genes and gene classes in a cogent fashion.

Barry et al. (2005) were the first to apply permutation testing to infer enrichment of gene classes. A statistic is computed for each gene class, defined abstractly, that depends on the differential expression of genes assigned to each gene class. The sample labels are permuted to derive a null distribution for the gene class statistics, producing a false discovery rate of enrichment for each class. Van Der Laan and Bryan (2001) proposed the parametric bootstrap, to perform gene subset selection accounting for covariance between the genes. The gene subsets can be defined in a general way, using gene class annotations. Lu et al. (2005) provide an in-depth discussion of multivariate gene detection with

Hotelling's T^2 statistic. Lu's approach is not designed specifically for inference of gene classes, although it generalizes well for the analysis of gene classes. Pan (2006) performed stratified gene detection given GO biological process annotations. Geoman et al. (2004) derive a score test for inferring the association of a gene expression in groups of genes with a clinical outcome, where the outcome is assumed to be a linear function of gene expressions. Wei and Li (2007) derived a nonparametric regression model for pathway associations with gene expression data. Liao et al. (2003) developed network component analysis for learning about gene pathway structure in multivariate data.

More advanced modeling includes group gene detection, given known feature groups $\mathcal{G}_1, \mathcal{G}_2, \ldots, \mathcal{G}_K$. In contrast to the lasso, group norm penalties $\|\boldsymbol{\beta}_{\mathcal{G}_k}\| = (\boldsymbol{\beta}_{\mathcal{G}_k}^T \boldsymbol{\beta}_{\mathcal{G}_k})^{1/2}$ are specified over the coefficients for differential gene expression in group $\mathcal{G}_k$, $\boldsymbol{\beta}_{\mathcal{G}_k} = (\beta_g)_{g \in \mathcal{G}_k}, k = 1, \ldots, K$ (Yuan and Lin, 2006; Zhao and Yu, 2006). The group lasso penalized loss function $L(\boldsymbol{\beta})$ with K groups is

$$L(\boldsymbol{\beta}) = (\boldsymbol{\beta} - \hat{\boldsymbol{\beta}}_{\mathrm{OLS}})' X' X (\boldsymbol{\beta} - \hat{\boldsymbol{\beta}}_{\mathrm{OLS}}) + \lambda \Sigma_k \|\boldsymbol{\beta}_{\mathcal{G}_k}\|. \qquad (6.7)$$

The group norm can be generalized to include group penalty, with general penalized loss function,

$$L(\boldsymbol{\beta}) = (\boldsymbol{\beta} - \hat{\boldsymbol{\beta}}_{\mathrm{OLS}})' X' X (\boldsymbol{\beta} - \hat{\boldsymbol{\beta}}_{\mathrm{OLS}}) + \lambda \sum_{k=1}^{K} \left(\|\hat{\boldsymbol{\beta}}_{\mathcal{G}_k}\|_{\gamma_k}^{\gamma_k} \right)^{\gamma_0}, \qquad (6.8)$$

for $\|\boldsymbol{\beta}_{\mathcal{G}_k}\|_{\gamma_k}^{\gamma_k} = (\Sigma_{g \in \mathcal{G}_k} |\beta_g|^{\gamma_k})^{1/\gamma_k}$. Different values of γ_k and γ_0 furnish different degrees of sparsity, either shrinking the coefficients of all variables in a group simultaneously to zero, or allowing some nonzero coefficients, as well as controlling the ordering by which genes are detected. Ma et al. (2007) applied the group lasso to gene expression array data.

From a Bayesian point of view, the group lasso solution can be expressed as the Bayes rule with zero–one loss, Gaussian likelihood, and prior

$$P(\boldsymbol{\beta}) \propto C_{\lambda, \gamma_0, \gamma} \exp \left\{ \lambda \sum_{k=1}^{K} \left(\|\boldsymbol{\beta}_{\mathcal{G}_k}\|_{\gamma_k}^{\gamma_k} \right)^{\gamma_0} \right\} \qquad (6.9)$$

on $\boldsymbol{\beta}$.

Example 6.2 Breast cancer case study
The breast cancer array data set of Farmer et al. (2005) was downloaded from http://www.ncbi.org/GEO. The R package GSA (Gene Set Analysis), which can be installed using the R graphical user interface in the usual way, was applied to detect important GO biological processes, contrasting gene expression in a subset of the subjects, with basal versus luminal breast cancers. GO gene classes with at least 50 and no more than 300 genes were included in the analysis, resulting in 194 gene classes, while reducing the number of genes to 10 825.

GSA was applied with a two-class unpaired comparison, using the maxmean gene class statistic and restandardization, for 200 permutation iterations. The package reports two permutation p-values for each gene class, for up and down regulation. For our purposes the GSA p-value p_{GSA} was evaluated as 2 times the minimum of the p-values for up and down regulation. Additionally, for comparison, the statistic λ_{max} was computed as the largest value of the penalty parameter λ in (6.7), such that the gene-wise parameters for change are not simultaneously zero, using the grplasso package in R. The λ_{max} statistic was derived for each gene class separately, with the responses for each gene weighted by an estimate of the respective gene-wise pooled scale. Larger values of λ_{max} indicate a greater degree of differential gene expression in the class, although they can also inflate in a locally linear fashion with number of genes in a gene class. The permutation distribution of the λ_{max}^*s was derived with 300 permutations of the subject labels. The λ_{max}s and permuted λ_{max}^*s were detrended separately by GO gene class gene counts, with ordinary least squares linear regression. Group lasso permuted p-values were obtained, after standardizing the detrended λ_{max}s and λ_{max}^*s, respectively, by mean and standard deviation, following the restandardization procedure of Efron and Tibshirani.

The empirical p-value distributions for both the GSA and group lasso applications are shown in Figure 6.1. Neither distribution shows any obvious departure from the uniform null distribution. At a liberal 0.05 p-value level, GO gene classes were selected: 13 GO classes by GSA and 10 by group lasso, with four GO classes discovered by both, printed in bold face in Table 6.3.

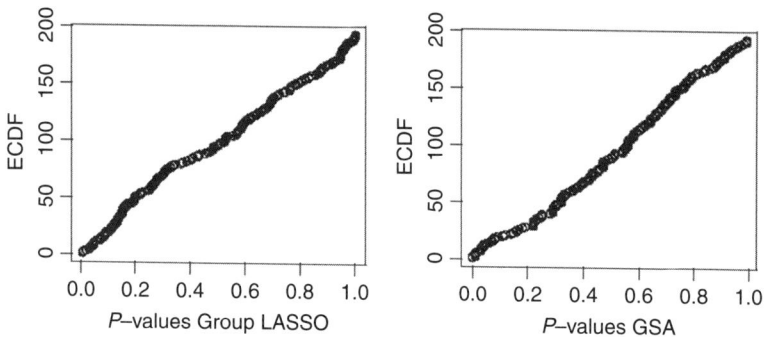

Figure 6.1   Gene set detection, Farmer breast cancer data, empirical p-value distributions: group lasso (left) and GSA (right).

6.4.1   Extending the Bayesian ANOVA Model

Enrichment analysis can be a very powerful method for inferring biological variation among gene classes, although it is a heuristic approach, first selecting a list of candidate genes given some experimental factors of interest, and then,

Table 6.3   Gene set detection: Farmer breast cancer data

Group lasso	GSA
1. **androgen receptor signaling pathway**	1. **androgen receptor signaling pathway**
2. **chromatin assembly or disassembly**	2. calcium ion homeostasis
3. **chromatin modification**	3. **chromatin assembly or disassembly**
4. epidermis development	4. **chromatin modification**
5. fatty acid metabolic process	5. glycolysis
6. **positive regulation of transcription, DNA-dependent**	6. negative regulation of apoptosis
	7. nucleotide biosynthetic process
7. positive regulation of transcription from RNA polymerase II promoter	8. nucleotide metabolic process
8. protein amino acid dephosphorylation	9. one-carbon compound metabolic process
9. regulation of Rho protein signal transduction	10. **positive regulation of transcription, DNA-dependent**
10. Wnt receptor signaling pathway	11. protein metabolic process
	12. regulation of Rab GTPase activity
	13. regulation of small GTPase mediated signal transduction

conditional upon that list of genes, choosing a set of gene classes. Information about GO biological processes or historical pathways is ignored during gene detection. While there is uncertainty in prior knowledge, integrating historical pathways with array data analysis is something that investigators do naturally. In this section, we discuss a Bayesian multistage linear model for use with microarray data, that incorporates prior information into the analysis. Unlike the ordinary linear model described in Chapter 3, an additional layer is included in the model, essentially to borrow strength between genes.

The multistage linear model, with two stages, can be expressed as

$$\mathbf{y} = X\boldsymbol{\beta} + \boldsymbol{\epsilon},$$
$$\boldsymbol{\beta} = Z\boldsymbol{\theta} + \mathbf{w},$$
$$\boldsymbol{\theta} = \boldsymbol{\theta}_0 + \mathbf{u}, \tag{6.10}$$

where **y** is a response of gene expression assumed to depend linearly on the unknown parameters for change β through the known matrix X, an experimental design matrix, of for example zero–one variables, as in ANOVA. The residual vector ϵ is distributed as multivariate Gaussian noise with mean 0 and variance covariance matrix Σ. The vector of unknown coefficients β determines the magnitude of the experimental effects explaining variation in the response **y**. These variables are linearly related to the vector θ through the (partially) known association matrix Z. The matrix Z is defined by additional knowledge in order to form associations between genes (rows) and gene classes (columns), discussed more fully below. The vector of residuals **w** is assumed to follow multivariate Gaussian noise with mean 0 and variance covariance matrix Σ_w. A multivariate normal prior is specified for θ as $MVN(\theta_0, \Sigma_\theta)$. Information from the top stage concerning **y** filters down to the next stage, accounted for by variation in β, which in turn is filtered down to the deepest stage, accounted for by variation in θ.

The multistage linear model can be made identifiable by including prior information about the association matrix Z, and sign constraints on Θ. Suppose that we truncate Θ below by $\delta > 0$, a reasonable prior guess at a moderate degree of differential expression considered biologically meaningful. Assume that the (g, c)th element of Z is a priori distributed as

$$
z_{gc} = \begin{cases} -a_g, & \text{with probability } \pi_1, \\ 0, & \text{with probability } \pi_2, \\ a_g, & \text{with probability } \pi_3, \end{cases} \tag{6.11}
$$

for $a_{gc} = 0$ if gene g is not a member of class b, and $a_{gc} = 1/\kappa_g$ if gene g is a member of gene class b, where κ_g equals the number of gene classes of which gene g is a member. Genes that are associated, as depicted by Z, are not necessarily expected to show co-expression. In a related context, this is accomplished by Bayesian mixture modeling, where the elements of Z identify the mixture components. Distinguishing the effects of multiple pathways on the gene expression of a single gene g is not feasible, without strong prior information. During Markov Chain Monte Carlo (MCMC) updating, at iteration c, if gene g is updated to be in a state of differential expression, all of the respective elements of Z associated with pathways, i.e. gene classes, for which gene g is a member, are updated with non-zero values. See Appendix B for an introduction to Bayesian Computational Tools.

The motivation for the group lasso, and group feature selection algorithms like it, lies in the strictly sparse solutions that these algorithms provide. Considering the risk, these algorithms furnish a shrinking effect, shrinking the unknown parameter estimates that are close to zero to zero, with the possibility of shrinking coefficients of features simultaneously to zero in the same group. While the multistage model does not enforce sparse solutions, it does shrink coefficient vectors in a convenient way. Genes that are not differentially expressed tend to have expression coefficients shrunk to zero, while differentially expressed genes tend to have coefficients for change that are shrunk to $Z\theta$. The bth component of the random variable θ, θ_b, represents the degree of differential expression in gene class b. If genes in gene class b exhibit moderate and consistent changes, for example, then θ_b will tend to be moderate, albeit with high precision. Unlike the previous models discussed, both a shrinking effect to the null, and a borrowing effect between differentially expressed genes, are possible with the multistage linear model.

Example 6.3 SCC Study

The Wachi data from case study 3.6 were analyzed with the multistage model to infer enrichment of GO biological processes. The goal is to learn if the regulation of any important biological functions are associated with SSC GO biological process classes with at least 50 and no more than 200 genes on the array were included in the analysis, resulting in 116 gene classes and 7748 genes. Log gene expression for each gene was mean centered, $\tilde{y}_g = y_g - 1_n \bar{y}_g$, modeled as

$$\tilde{y}_g = X\beta_g + \epsilon_g,$$

$$\epsilon_g \sim N(0, \sigma_g^2 I_n),$$

$$\sigma_g^{-2} \sim \text{Gamma}(a_0/2, b_0/2), \tag{6.12}$$

for each gene, with $X = 1_5 \otimes (-1, 1)^T$. A priori, the vector $\beta = (\beta_1, \beta_2, \ldots)$ was modeled as

$$\beta \sim N(Z\theta, W\Sigma) \tag{6.13}$$

with Z the 7748×116 stochastic association matrix, defined as above, and $\Sigma = \text{diag}(\sigma_1^2, \ldots, \sigma_G^2)$ for user-defined W. The prior distributions for the z_{gc} were defined as in (6), with $\pi_1 = \pi_2 = \pi_3 = 1/3$. The parameter θ is distributed a priori as

$$\theta \sim N(1_{116}\theta_0, \Omega)I(\theta > \delta), \tag{6.14}$$

with $\delta > 0.5$ reflecting the desire to detect changes in gene expression of magnitude at least 2 on the original scale. The diagonal hyperparameter matrix $\Omega = \tau \text{diag}(\kappa_1^{-1}, \kappa_2^{-1}, \ldots)$ is defined by a user constant τ divided by the number of genes in each gene class, along the main diagonal. Several combinations of hyperparameters, $W = 1, 3, 5, 7$ and $\tau = 0.05, 0.1$, were fitted to the data, indexing different models $m_1, \ldots, m_8$. The best combination of W and τ was chosen by the Bayesian model selection criterion, maximizing $\log P(Y|m_j)$ (Figures 6.2 and 6.3).

Having reduced the number of gene classes to 116 and the number of genes to 7,748, the remaining genes were used to determine the hyperparameters a_0, b_0 and θ_0. The median sample variance among the remaining genes was 0.0357; hyperparameters were set to $a_0 = 5$ and $b_0 = 0.15$, reflecting moderately strong beliefs. The sample estimate of the absolute mean change $|\beta|$ in gene expression, given that $|\beta| > 0.5$, was 0.77, and therefore we set $\theta_0 = 0.77$.

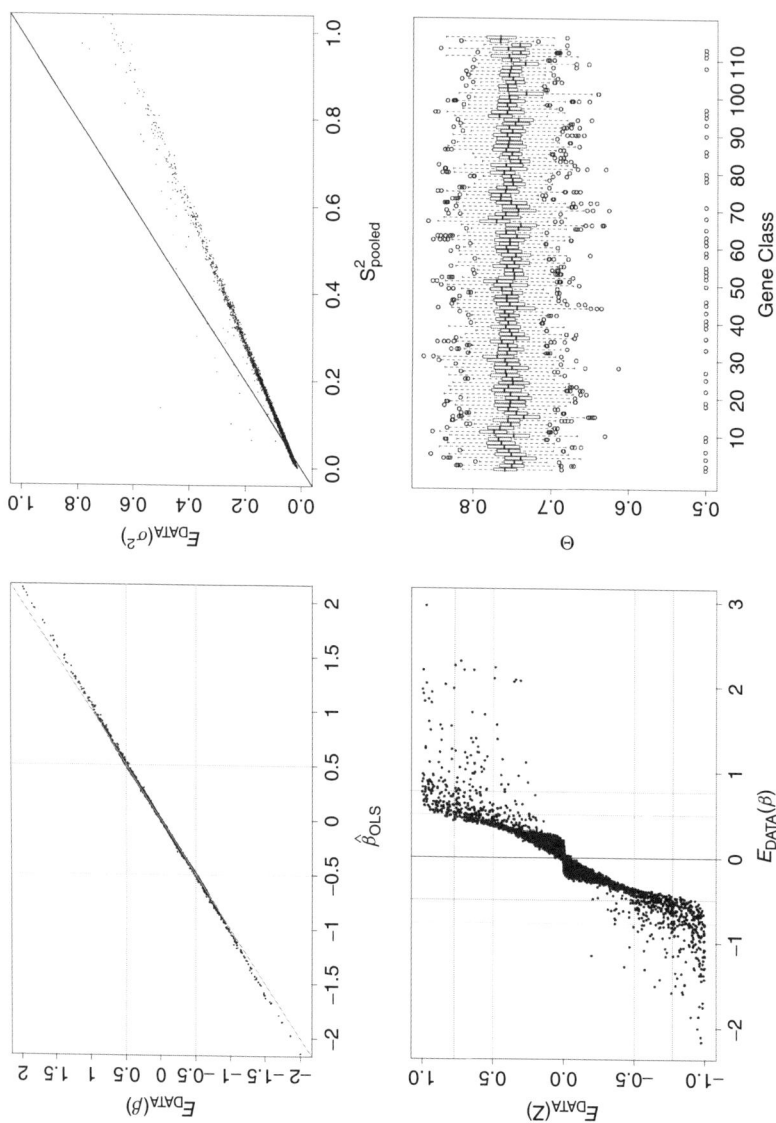

Figure 6.2   Enrichment Analysis of the Wachi Experiment with Bayesian Multistage Linear Model Analysis, $W = 7$ and $\tau = 0.05$.

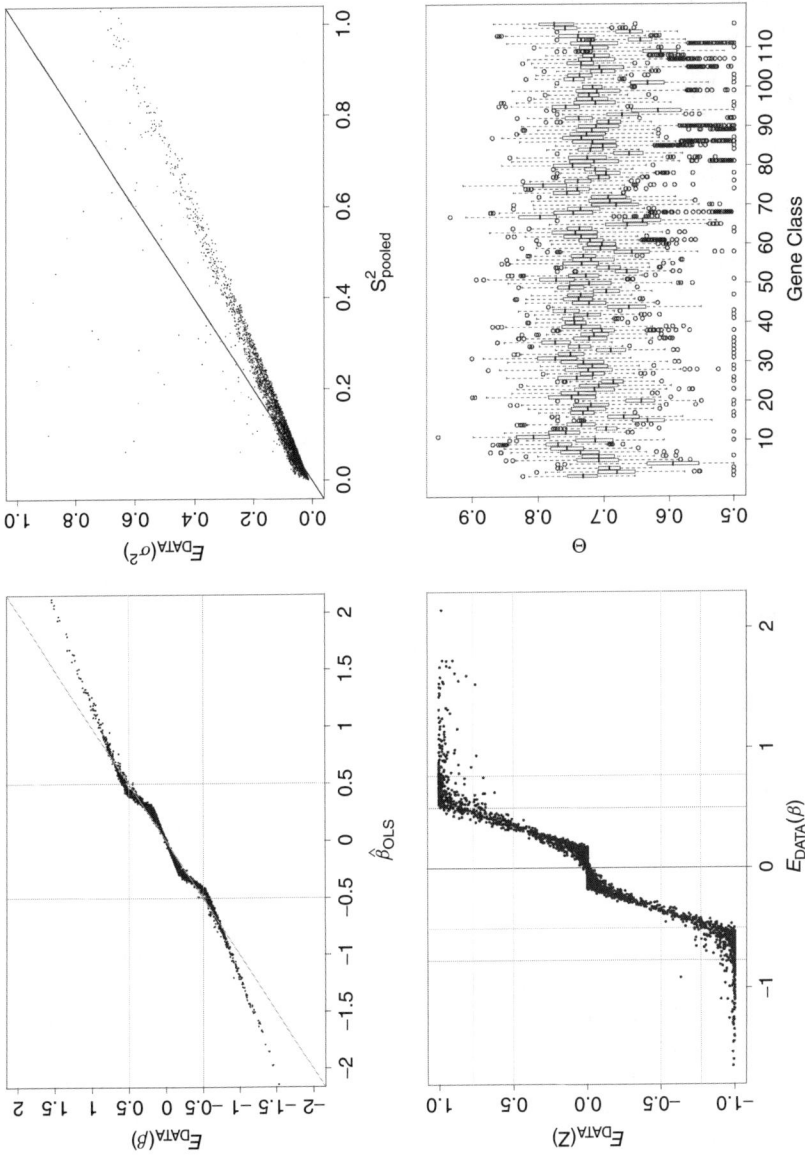

Figure 6.3   Enrichment Analysis of the Wachi Experiment with Bayesian Multistage Linear Model Analysis, $W = 1$ and $\tau = 0.1$.

Table 6.4   Bayesian multistage enrichment analysis, Wachi data

Gene class	No. of genes.	Posterior expectation Υ_b (2)
transmembrane receptor activity	154	0.998
sugar binding	178	0.998
structural constituent of cytoskeleton	155	0.998
serine-type endopeptidase inhibitor activity	89	0.998
ribonuclease activity	71	0.998
protein N-terminus binding	62	0.998
protein kinase inhibitor activity	50	0.998
protein kinase binding	80	0.998
protein heterodimerization activity	170	0.998
protein binding, bridging	52	0.998
phosphoric monoester hydrolase activity	63	0.998
oxygen binding	59	0.998
nuclease activity	149	0.998
NAD binding	56	0.998
monooxygenase activity	117	0.998
kinase inhibitor activity	52	0.998
kinase binding	114	0.998
integrin binding	85	0.998
heparin binding	129	0.998
heme binding	161	0.998
GPI anchor binding	147	0.998
endopeptidase inhibitor activity	122	0.998
endonuclease activity	69	0.998
electron carrier activity	132	0.998
copper ion binding	81	0.998

The combination $W = 1$ and $\tau = 0.05$ was chosen to maximize $\log P(Y|m_j)$, found by Monte Carlo integration. These hyperparameters reflect a relatively informative prior for θ and moderate to weak dependence between β coefficients for the genes (Figure 6.4). Contrast these results with the combination $W = 7$ and $\tau = 0.1$, allowing a weaker prior for β, while inducing more dependence between the genes (Figure 6.4). Notice that in both cases some shrinkage of the parameter estimates from the OLS estimates is induced, with more shrinkage at $W = 7$, while the larger gene-wise parameters for variance are shrunk down from their pooled estimates.

The top 25 most highly enriched GO classes, ordered by $E(\Upsilon)$ (6.2) are listed in Table 6.4. Out of the top 25 functions, 14 functions are shared with the univariate analysis in Example 11.1.

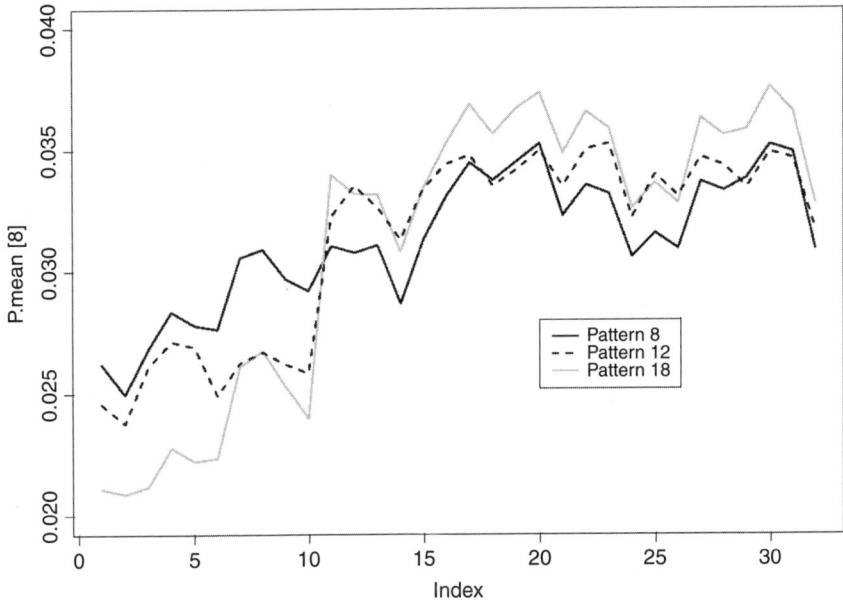

Figure 6.4   BD analysis of Klevecz yeast time course data: three enriched patterns.

6.4.2   Bayesian Decomposition

Enrichment analysis can be a very useful application for learning about the association of important pathways with phenotypes in designed controlled experiments. In some cases, the design of an experiment or case–control study does not lend itself well to or permit direct inference on a factor of interest. For example, changes in gene expression over time in a time course gene expression study can be very difficult to relate to a particular pathway or function since there are so many possible combinations of ways genes can vary in expression over multiple time points. In contrast to controlled factorial designs, associating pathways with significant temporal patterns is a tenuous task. We discuss Bayesian methods for analysis of time course gene expression studies in Chapter 9, and here demonstrate the use of a novel Bayesian pattern recognition method proposed for exploring pathway associations with temporal gene expression patterns.

Moloshok et al. (2002) proposed Bayesian decomposition (BD), a pattern recognition algorithm that allow genes to cluster, and borrow strength, in multiple expression patterns. The fundamental idea is that an $G \times n$ expression matrix X,

$$X = A\Gamma$$

$$
\begin{pmatrix}
x_{11} & x_{12} & \cdots \\
& & \\
\vdots & & \\
& & \\
& x_{Gn} &
\end{pmatrix}
=
\begin{pmatrix}
a_{11} & \cdots \\
& \\
\vdots & \\
& \\
& a_{np}
\end{pmatrix}
\begin{pmatrix}
\gamma_{11} & \gamma_{12} & \cdots \\
& \ddots & \\
& & \gamma_{pn}
\end{pmatrix}, \quad (6.15)
$$

can be decomposed into a $p \times n$ matrix of expression patterns Γ, and a matrix of $G \times p$ amplitudes A. Each row of Γ corresponds to a 'common' expression pattern found among genes in X, while each row g in A provides the amplitudes, or weights, of association between gene expression profile g and the patterns in Γ. For example, the largest amplitude in the gth row of A, $\max(A_{g1}, A_{g2}, \ldots, A_{gp})$, corresponds to the dominant pattern found in Γ associated with the expression profile of gene g. The second largest amplitude corresponds to the next most dominant pattern in Γ associated with expression profile of gene g, and so on. Moloshok et al. (2002) put prior distributions on the matrices Γ and A, with the the number of patterns p defined explicitly. One may assign informative priors for the patterns in Γ, as well as dependent joint priors linking the rows of A, based on prior pathway information. The prior distribution on A can specify the joint probability that any collection of genes will co-express across one or more of the patterns:

$$P(A, \Gamma | X) \propto P(X | A, \Gamma) P(A, \Gamma). \quad (6.16)$$

Multiple solutions of the decomposition (6.15), can exist that fit the data well. For this, a search is conducted to compare multiple solutions, to sample the ones that are more probable. Note that while assignment of informative priors is feasible with Bayesian decomposition, in practice it can be cumbersome and awkward to translate prior information into the expression domain.

An advantage of Bayesian decomposition is the utility to associate the posterior expression patterns discovered with genes among gene classes. Enrichment analysis is performed, updating the random variable

$$\Upsilon_b^p = I(\|\mathcal{G}_p \cup \mathcal{G}_b\| / \|\mathcal{G}_p\| \geq \|\mathcal{G}_p^c \cup \mathcal{G}_b\| / \|\mathcal{G}_p^c\|), \quad (6.17)$$

contrasting each pattern p, by class b, via a selection criterion cutoff applied to the weights in A.

Example 6.4 Yeast time course study
In the final example of this chapter, we explore associations between complex temporal gene expression profiles and known gene transcription regulators in a

yeast time course study. The software BD (http://bioinformatics.fccc.edu) implements Bayesian decomposition, allowing for user specification of the standard deviation of the noise in the data, σ, assumed to be the same for all genes, the number of patterns, or rows in Γ, and the number of iterations for convergence. The algorithm provides the posterior expectation and posterior standard deviation of the data predictions, pattern matrix P, and amplitude matrix A. It also provides the log of the marginal predictive distribution of the data $\log P(X)$ integrating over Γ and A.

The Klevecz et al. (2004) Affymetrix S98 yeast arrays were downloaded, including 32 time points over three cycles of dissolved oxygen. Yeast transcription factors and their downstream targets were obtained from http://www.yeastextract.org, defining 17 gene classes, including 703 targets.

BD was run for combinations of $p = 5, 10, 15, 20, 25$ patterns and uncertainty $\sigma = 0.4, 0.3, 0.2, 0.1, 0.05$ for 1000 iterations each. Utilizing the Bayesian model selection criterion, the model with $p* = 20$ patterns and $\sigma* = 0.4$ was selected, i.e. maximizing the log predictive distribution $\log P(X)$.

Three transcription factors (classes) showed a large portion of respective downstream targets (gene members), following the analysis of Moloshok et al. (2002), with at least 70% of variation explained by the posterior expectation of a pattern in $P|X$. These were HAA1, with 40% of genes associated with pattern 12, MET31 with 41% of genes associated with pattern 18, and OPI1, with 50% of genes associated with pattern 8, displayed in Figure 6.4.

6.5   Summary

Inferring differential expression among a priori gene classes, associated with a phenotype or experimental factor of interest, remains an open research question, and will remain so, as no one approach is necessarily correct for every analysis. There are many ways to define the hypothesis of change for a gene class, with enrichment being the convention. The heuristic enrichment analysis approach considers gene and gene class detection separately. This is a reasonable approach to take, and in practice is a sensible first step. More advanced frequentist methods, such as GSA, show promise, providing for statistical considerations.

Incorporating gene dependence as in the multistage model or through prior specification with Bayesian decomposition should not be regarded as replacing but rather as supplementing a one-at-a-time gene analysis. The reason for this is that there is still much uncertainty surrounding gene annotations, and potential uses of the prior information. The goals of gene expression analyses and and ongoing efforts to genotype diseases do not necessarily have to be independent aims, as efforts to integrate learning between a spectrum of studies provide new incentives and directions. The Bayesian learning paradigm is a natural one to adopt, although any analysis aimed at using prior information should largely reflect the fact that, in the state of post-genomic research, the devil, so to speak, is in the detail.

Ever more exciting and challenging statistical research will no doubt improve the way we can pool information between genes, and infer important patterns in known pathways, to give researchers focus and direction toward historical pathways and gene targets. Despite the uncertainty, there can be more to learn from high-throughput experiments, if we are careful to consider the functional roles of genes and their course in evolution.

7

Unsupervised Classification and Bayesian Clustering

7.1 Introduction to Bayesian Clustering for Gene Expression Data

In Chapter 5 we introduced the problem of Bayesian supervised classification for gene expression data, for example, to predict patient disease status, given gene expression profiles and clinical covariates. For pairs of observed responses and gene expression profiles $(y_i, \mathbf{x}_i)$ for individuals $i = 1, 2, 3, \ldots$, a classifier, or prediction, rule is constructed with the goal of predicting the response of a new individual given gene expression profiles $\mathbf{x}_{\text{new}}$. The goals of supervised classification can be several-fold, although the utility of the predictor is measured by how well it can discriminate, for example, pathological status or response to therapy, given observed gene expression profiles.

Some diseases such as cancer are moxlecularly heterogeneous, appearing to produce similar symptoms and cellular irregularities, although very different responses to therapy. Many of the important differences in gene expression between diseases have yet to be discovered. It is for this reason that we turn to a different although related problem: unsupervised classification, also commonly referred to as cluster analysis (we use the two terms interchangeably throughout). Suppose that we observe gene expression $\mathbf{x}_i$ for individuals $i = 1, 2, 3, \ldots$, originating from $k = 1, \ldots, K$ biological classes where the class membership y_i for each individual as well as the number of classes C are unknown – hence the name unsupervised classification, as there is no a priori class knowledge to guide learning. In principle, individuals with similar expression profiles are more likely to belong to the same cluster, while individuals with very different gene expression should belong to different clusters. Clustering algorithms based on this principle assign individuals to clusters by a measure of similarity in observed gene expression.

In this chapter, our aim is to predict class membership of individuals, where the individuals can be the subjects (e.g. patient disease status) or the genes (e.g. to discover clusters of highly associated genes), or perhaps to cluster both the subjects and the genes. Some of the methodologies that we present in this chapter lend themselves more conveniently to clustering just the subjects or genes, requiring constraints or modification to cluster both. In order not to confuse these two important, though not necessarily competing, aims, we make distinctions throughout this chapter where applicable between clustering the subjects and clustering the genes.

Example 7.1 Bi-way clustering of lung cancer cell lines
We illustrate the results of a common clustering task, bi-way clustering, of gene expression profiles with the Zhou et al. (2006) study of expression in 79 lung cell line samples from 13 origins, including adenocarcinomas, squamous cell carcinoma, mesothelioma and normal lung tissue. The Affymetrix U133A data were downloaded from the NCBI/GEO data retrieval system (http://www.ncbi.nlm.nih.gov/) and transformed by the natural logarithm, truncating values less than 0. Probes were filtered, including only probes with at least 50% of values above the overall data set median, resulting in 10 881 probes. The cell lines and genes were both clustered, each independently, with the techniques described in Section 15.3. Notice in Figure 7.1 the yellow (low) and red (high) 'expression patterns' that appear by cell line (rows) and genes (columns). These patterns can be indicative of pathological subtypes and gene clusters associated with disease.

Figure 7.1   Heatmap: Bi-way clustering of lung cancer cell lines, 79 samples (rows) and 10 881 probes (columns). High (red) and low (yellow) probe expression patterns distinguish the samples.

The results of the lung cancer study of Example 7.1 suggest that cancer-specific molecular signatures can be discovered and used to discriminate subjects. Keep in mind that the results and conclusions of applied cluster analysis with gene expression data can be hotly debated. Part of the reason for this is that classification, as we saw in Chapter 14, involves potential risks to classify subjects incorrectly, and the cost of making a wrong decision can be high. It is widely believed that the results from these studies should be validated with more data, or external means. The nature of this problem is further compounded when we do not know the true classification labels, i.e. the potential to read too much into the analysis or infer from apparent patterns – false signatures. With this in mind we proceed with clustering, and list the advantages that Bayesian clustering methods can deliver in a variety of situations.

Among the many popular approaches for unsupervised classification that we will discuss in this chapter are hierarchical clustering, principal component clustering, model-based clustering, K-means clustering, mixture model clustering, and Dirichlet (Chinese restaurant) process clustering. Bayesian generalizations of many of these clustering methods are well known and can provide advantages for learning. An important distinction though, from frequentist clustering methods, is that to a Bayesian there is uncertainty in cluster membership, both before seeing the data, and after. This fundamental distinction implies that membership in a cluster is only known up to a measure of uncertainty. We begin discussion of Bayesian clustering by first introducing the fundamentals from classical work in the field.

7.2   Hierarchical Clustering

Hierarchical clustering algorithms come in many forms. In this section we discuss one of the most popular forms, recursive binary clustering. No distinction is required in order to apply the algorithm to cluster the subjects or the genes, and the algorithm may be applied to cluster both, independently.

In order to perform the hierarchical clustering algorithm, rules are specified for measuring distances between individuals and between clusters, and the direction of agglomeration:

1. Define a measure of pairwise distance between individuals i and $j \neq i$.

2. Adopt a linkage rule.

3. Choose the direction of aggregation (top-down or bottom-up).

As in other clustering methods, measures of dissimilarity or distance between individuals are important. Some common measures of distance between multivariate vectors are defined in Table 7.1.

Table 7.1   Distance Metrics for hierarchical clustering

Name	Formula
Euclidean	$D_{\text{Euclid}}(\mathbf{x}_j, \mathbf{x}_k) = \sqrt{(x_{1j} - x_{1k})^2 + \ldots + (x_{pj} - x_{pk})^2}$
Correlation	$D_{\text{Corr}}(\mathbf{x}_j, \mathbf{x}_k) = 2 \times (1 - r_{X_j, X_k})$
Manhattan	$D_{\text{Block}}(\mathbf{x}_j, \mathbf{x}_k) = |x_{1j} - x_{1k}| + \ldots + |x_{pj} - x_{pk}|$

Linkage defines how distance is measured between clusters, and between individuals and clusters. The three commonly used linkage methods are nearest neighbor, average and complete linkage. Consider two clusters $c = 1, 2$, consisting of n_1 and n_2 subjects. Let $D(X\mathbf{x}_i, \mathbf{x}_k)$ be the distance between the ith individual in cluster 1, and the kth individual in cluster 2. Nearest neighbor linkage defines the distance between any two clusters as the minimum of the $D(\mathbf{x}_i, \mathbf{x}_k)$, i.e. the shortest distance between between all pairs of individuals from the respective clusters. Average linkage defines the distance between two clusters as the sample average of the $D(\mathbf{x}_i, \mathbf{x}_k)$ between all pairs of individuals from the respective clusters. Complete linkage defines the distance between two clusters as the maximum of the $D(\mathbf{x}_i, \mathbf{x}_k)$, i.e. the farthest distance between all pairs of individuals from the respective clusters.

Methods for agglomeration can either be bottom-up, or top-down. In bottom-up agglomeration all points are initially in their own own cluster. The first cluster is formed by combining the pair of individuals that are closest in terms of distance. At the next step, the individual(s) and or cluster with the shortest distance are merged into a cluster. The process continues until all individuals are combined into one cluster. In top-down agglomeration all individuals are initially assumed to belong to one cluster. The cluster is split into two groups so that the measure of distance between the clusters is greatest. At the next step, one of the respective clusters is split, so that the two new clusters that are formed are the farthest apart in terms of distance. This process continues until all individuals belong to only one cluster.

A graphic commonly used for visualizing hierarchical cluster results is the **dendrogram**. On the horizontal axis all individuals are displayed, ordered by their respective cluster assignments. Above and between each individual/cluster is a single branching node, connected to two branches below and one above. Nodes are positioned at the height equal to the distance of the merged clusters. Figure 7.2 shows a dendrogram for the lung cell line data of Example 7.1. In order to assign cluster membership, a cutoff is chosen along the vertical axis. Individuals that are merged by a branching node immediately above the cutoff height are assigned to the same cluster.

7.3 K-Means Clustering

Consider a population with K p-dimensional sub-populations or clusters, of genes or subjects, as clustering of either is possible with K-means. Each cluster is assumed to have its own unique density with mean $\mu_k = (\mu_{1k}, \ldots, \mu_{kp})$. More generally, μ_k is considered to be the 'center of mass' or 'center of gravity' to which points aggregate in the kth cluster. The essential feature of the algorithm is to choose $\{\hat{\mu}_1, \ldots, \hat{\mu}_K\}$ and $\{A_1, \ldots, A_K\}$ in order to minimize the score

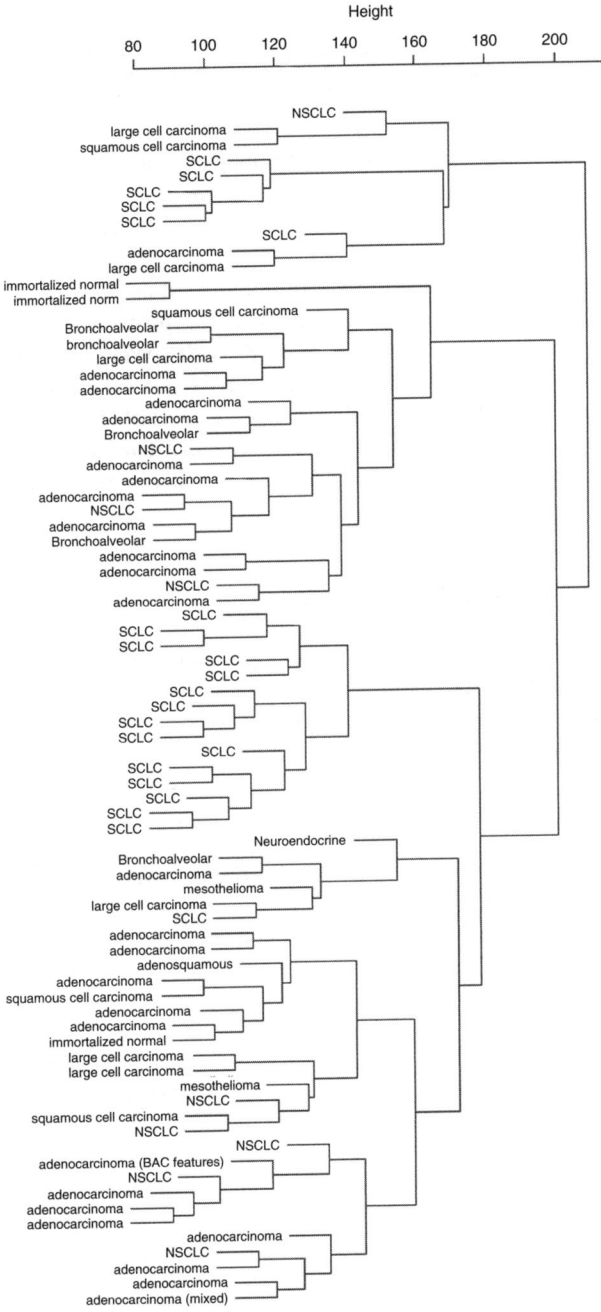

Figure 7.2 *Dendrogram for lung cell line data.*

function

$$S(\boldsymbol{\mu}; X) = \sum_{k=1}^{K} \sum_{i \in A_k} \sum_{g=1}^{p} \left(x_{gi} - \mu_{gk}\right)^2, \tag{7.1}$$

where the set $A_k = \{i : \text{subject } i \text{ is a member of cluster k}\}$. Maximizing the K-means score function is synonymous with maximizing the product of K p-dimensional Gaussian likelihoods, with diagonal covariance matrix of equal dispersion along the main diagonal.

In practice, choosing K is a difficult problem, as it defines how many clusters are in the data. Frequentist methods deal with the problem of choosing K by assessing the fitness of different values of K with resampling-based methods, such as cross-validation and the bootstrap. For a fixed K, the Bayesian extension of the K-means algorithm is straightforward, as described in the next section on model-based clustering. In addition, the Bayesian can specify a prior on K, the number of clusters, allowing integration over all of the uncertainty in K in determining cluster membership.

7.4   Model-Based Clustering

Consider a set of K clusters, in p-dimensional space, each distributed according to parametric densities $P_k(\theta_k)$, for $k = 1, \ldots, K$. If the densities P_k were known then the individuals, subjects or genes, could be assigned to the cluster maximizing the respective likelihood

$$k(i) = \text{argmax}_k P_k(\mathbf{x}_i; \theta_k), \tag{7.2}$$

for $k = 1, \ldots, K$. In practice, the number and distributions of the cluster populations are unknown. Suppose that the K clusters are distributed according to the multivariate Gaussian distribution. Conditioning upon cluster label $y_i = k$, the likelihood of the p-dimensional variable $\mathbf{x}_i$ is

$$\mathbf{x}_i | y_i = k \sim N_p(\boldsymbol{\mu}_k, \Sigma), \tag{7.3}$$

indexed by unknown mean vector μ_k and unknown variance–covariance matrix Σ. Here the aim is to determine the number of cluster populations K, estimate the unknown means and covariance, and assign cluster membership. This can be achieved by choosing $\hat{\mu}_k$, $\hat{\Sigma}$, $\hat{y}_i$ and $\hat{K}$ to maximize the likelihood

$$P(\mathbf{y}|\boldsymbol{\mu}, \Sigma, X) = \prod_{i=1}^{n} \prod_{k=1}^{\hat{K}} N(\mathbf{x}_i; \hat{\mu}_k, \hat{\Sigma})^{I(y_i=k)}. \tag{7.4}$$

A heavy-tailed alternative to the multivariate Gaussian distribution is the multivariate t distribution: see the R package mclust for details and model fitting.

Note that like principal component clustering discussed in Section 7.7, clustering genes with full off-diagonal covariance matrix Σ is typically not possible without simplifying assumptions, due to the number of subject samples required. Model-based clustering of genes is made possible by, for instance, constraining the off-diagonal elements of Σ to be 0.

Extending the model-based clustering framework to Bayesian clustering is straightforward. Posterior inference proceeds by specifying prior distributions for unknown parameters. Suppose, for example, that $\mathbf{x}_i$ is distributed conditionally as

$$\mathbf{x}_i | y_i = k \sim N(\mu_k, \Sigma), \tag{7.5}$$

with p-dimensional mean μ_k and $p \times p$ diagonal covariance matrix $\Sigma = \text{diag}(\sigma_1^2, \dots, \sigma_p^2)$. We specify prior distributions

$$\mu_k \sim N_p(\mathbf{m}_k, V),$$

$$\sigma_g^2 \sim IG(a_g, b_g),$$

$$y_i \sim \text{Multinomial}(\mathbf{1}, \pi) \tag{7.6}$$

Integrating out the parameters, the posterior probability of $y_i = k | \mathbf{x}_i$ is

$$P(y_i = k | \mathbf{x}_i) \propto \pi_k \iint N(\mathbf{x}_i; \mu_k, \Sigma) N(\mu_k; \mathbf{m}_c, v^2) \prod_{g=1}^{p} IG(\sigma_g^2; a_g, b_g) d\sigma_g^2 d\mu_k$$

$$\propto \pi_k S(\mathbf{x}_i; m_k, Q), \tag{7.7}$$

for $p < G$, i.e. proportional to a multivariate t distribution with location m_k and dispersion matrix Q. While the frequentist and Bayesian criteria appear identical, recall that only Bayesians can make probability statements about the parameters, i.e. y_i, given the data.

7.5 Model-Based Agglomerative Hierarchical Clustering

A disadvantage of model-based clustering is the computation time required to evaluate the likelihood for all possible clusters. Combining the computational efficiency of hierarchical clustering with the model-based approach of model-based clustering leads to model-based agglomerative hierarchical clustering. The idea is to consider the likelihood as a score function to be optimized. The likelihood is given by

$$P(\mathbf{y}|\theta_1, \dots \theta_K; l_1, \dots, l_n) = \prod_{i=1}^{N} P_{l_i}(y_i|\theta_{l_i}), \tag{7.8}$$

where the l_i are the group labels. The likelihood is maximized in stages, merging the pair of clusters achieving the greatest increase at each stage. A disadvantage of model-based agglomerative hierarchical clustering is that at each stage the results are conditional upon the previous stage. The algorithm does run much faster than model-based clustering, and can lead to similar results.

7.6 Bayesian Clustering

The landmark paper in Bayesian clustering is that of Binder (1978) who first introduced the theoretical innovations for Bayesian clustering. Let X denote an $n \times p$ matrix of p measured characteristics for each of n individuals, i.e. $\mathbf{x}_i = \{x_{i1}, \ldots, x_{ip}\}$ denotes the measured characteristics for the ith individual. Let $\mathbf{y} = (y_1, \ldots, y_n)$ denote the (at least partially) unobserved cluster membership of each individual i, i.e. $y_i = k$ denotes that the ith individual is a member of the kth cluster, with K the (unknown) total number of clusters. Without loss of generality, Binder's method can be applied to cluster subjects, in which case the $\mathbf{x}_i$ are the measured expression profiles of each subject across a collection of genes on the array, or to cluster the genes given the subjects profiled across the study, in which case the response is indexed by $g = 1, \ldots, G$. We omit any distinction between clustering subjects and genes for the present discussion, considering individuals to be appropriately defined for the aim of the analysis. Individuals are assumed conditionally independent, given $\mathbf{y}$,

$$\mathbf{x}_i | y_i = k, \theta \quad i.i.d. \ P_k(\mathbf{x}_i | \theta_k), \tag{7.9}$$

with density $P_k(\mathbf{x}_i | \theta_k)$. The prior for the unknown variables $(K, \mathbf{y}, \theta)$ is

$$P(K, \mathbf{y}, \theta) = P(K)P(\mathbf{y}|K)P(\theta|K, \mathbf{y}). \tag{7.10}$$

Notice that the prior density for θ is defined conditionally upon $\mathbf{y}$. In the case where $P(y_i = k|K) = \pi_k$, introducing the unobserved prior weights for cluster membership, the prior is

$$P(K, \mathbf{y}, \boldsymbol{\pi}, \theta) = P(K)P(\boldsymbol{\pi}|K) \prod_{k=1}^{K} \pi_k^{n_k} P_k(\theta_k|\mathbf{y}), \tag{7.11}$$

where n_k is the number of members in the kth cluster.

Suppose further that θ does not depend on y, in practice a reasonable and convenient assumption to make. Then the likelihood is

$$P(X|K, \boldsymbol{\pi}, \theta) = \sum_{k=1}^{K} \pi_k P_k(X|\theta_k), \tag{7.12}$$

i.e. a mixture over P_k with mixture weights π_k for $k = 1, \ldots, K$. The marginal posterior $\mathbf{y}|X$ is given by

$$p(\mathbf{y}|X) \propto \sum_K p(K)p(\mathbf{y}|K) \times \prod_{k=1}^{K} \int_{\Theta_c} P(\theta_k|K, \mathbf{y}) \cdot \prod_{i \in A_k(\mathbf{y})} P_k(\mathbf{x}_i|\theta_k)d\theta_k \quad (7.13)$$

where the set $A_k(\mathbf{y}) = \{i : y_i = k\}$ includes the indices of samples assigned to cluster k, and the integral is over θ_k where $P(\theta_k|K, A_k(\mathbf{y})) > 0$. An advantage of Bayesian clustering is that it provides a probability measure of uncertainty in class membership. The marginal prior probability $P(y_i = k)$ is updated in a straightforward way to yield $P(y_i = k|X)$, a measure of posterior precision. In contrast to frequentist clustering, y_i is not assumed fixed. In the event that a point prediction is desired, Bayesian clustering rules include criteria for:

1. maximizing the joint posterior probability;

2. for each individual i, choosing the group $k = \text{argmax } P(y_i = k|X)$, i.e. the marginal posterior probability;

3. performing an available clustering algorithm, defining pairwise similarity as $P(y_i = y_j|X)$;

4. minimizing a posterior risk function.

Evaluating $P(\mathbf{y}|X)$ for all possible permutations of the cluster labels $y = \{1, 2, 3, \ldots\}$ is computationally cumbersome. For K clusters and n individuals there are K^n possible ways to cluster individuals, allowing for empty clusters. An approximation to the posterior mode can be found by choosing $\hat{y}_i$, for $i = 1, \ldots, n$, maximizing

$$S_* = \sum_{k=1}^{K} \sum_{i \in A_k} \sum_{l \in A_k} P(\hat{y}_k = \hat{y}_l|X) - \frac{1}{2} \sum_i \hat{n}_k^2 \quad (7.14)$$

for the set A_k of members in the kth cluster and $\hat{n}_k = |A_k|$ (Binder, 1981). This sum is maximized in two stages:

I Obtain a list of local maxima, by a hill climbing procedure, based on an approximation of S_*.

II Search the neighborhoods of local maxima and choose the partition with the largest value.

7.7 Principal Components

Consider the gene expression matrix X, with gene expression values in the rows and subject samples in the columns. The estimated covariance matrix between subject samples is

$$V = (N-1)^{-1}S = (N-1)^{-1}\sum_{i=1}^{N}(\mathbf{x}_i - \bar{X})(\mathbf{x}_i - \bar{X})^{T}. \qquad (7.15)$$

The singular value decomposition (SVD) of V is

$$V = EDE^{T}, \qquad (7.16)$$

where D is a diagonal matrix of eigenvalues $(d_1, d_2, \ldots, d_p)$ ordered from largest to smallest down the main diagonal, and E is the corresponding orthogonal matrix of eigenvectors, i.e. $E^{T}E = I$. The first principal component (PC) is defined as $P_1 = X^{T}E_1$, the second as $P_2 = X^{T}E_2$ etc., where E_j is the jth eigenvector corresponding to the jth largest eigenvalue, $d_j = \text{var}(P_j)$. Intuitively, E_1 captures the dominant profile in gene expression values, across subjects, accounting for the largest PC, E_2 the second largest PC, etc. PC clustering has been applied to explore the dominant sources of variation between array samples in gene expression data that may be attributable to biology or experimental variation. Unlike the previous methods, PC clustering requires a full-rank positive semi-definite sample covariance matrix V, and therefore it is 'typically' not possible to perform PC clustering on all of the genes on the array, given the subjects, without a substantial number of samples.

A scatter plot of the first two eigenvectors, E_2 versus E_1, provides a visual display of clusters discriminated the first two PC reductions of their expression profiles. Further scatter plots can be useful, of the first and third eigenvectors, the second and third, etc., to discover subject clusters.

How many PCs are needed to explain the major sources of variation in the data? Scree plots provide a visual approach to choosing the number of meaningful PCs. The eigenvalues are plotted from largest to smallest. An obvious break in the plot (Figure 7.3(d)) between the eigenvalues yields evidence of the maximum number of PCs needed. In practice, inference is desirable for choosing the number of components accounting for the major variation in the data. It has been shown that the frequentist asymptotic distribution of the eigenvalues is multivariate normal (Wigner, 1955; Dhesi and Jones, 1990). In a Bayesian PC analysis, one assumes that the true covariance matrix is unknown, and specifies a prior distribution for it. For a general variance–covariance matrix with nonzero off-diagonal terms, suppose we adopt an inverse Wishart prior on Σ,

$$\Sigma \sim \text{InverseWishart}(\alpha, R), \qquad (7.17)$$

with parameters α, the effective prior sample size, and matrix R, a prior guess at Σ. Suppose that the SVD of Σ is $E\Theta E^{T}$. The posterior is of Σ is inverse

Wishart

$$\Sigma | X \sim \text{InverseWishart}(\alpha*, R*),$$

$$\alpha* = \alpha + n,$$

$$R* = R + S, \tag{7.18}$$

with posterior parameters $\alpha*$ and $R*$ and posterior expectation $E(\Sigma|X) = (\alpha + N - p - 1)^{-1}(R + S)$. One can generate samples from the posterior of Σ, transform them by the SVD, and derive posterior probability statements about the θ_j.

Example 7.2 PC clustering of lung cancer cell lines
PC plots of the Lung Cell Line data (see Example 7.1) are shown in Figure 7.3. A strong contrast is observed in the second component. The scree plot in Figure 7.3(d) suggests that more than two factors separate the cell lines. A Bayesian PC analysis was applied (Figure 7.4). The 95% Bayesian posterior credible envelope fits very tightly around the eigenvalues. As many as six significant components can be inferred.

7.8 Mixture Modeling

Mixture modeling is an attractive way to identify clusters in the data. In this setup we assume that the multivariate gene expression profiles $x_1, \ldots, x_n$, genes or subjects, are distributed as a mixture with likelihood

$$P(X|\theta_1, \ldots \theta_K; \pi_1, \ldots, \pi_K,) = \prod_{i=1}^{n} \sum_{k=1}^{K} \pi_k P_k(x_i|\theta_k), \tag{7.19}$$

with P_k the multivariate Gaussian distribution, $0 \leq \pi_k \leq 1$ mixture weights, and $\theta_c = (\mu_c, \Sigma_c)$. The space of possible parameterizations of Σ is rather large, and must be appropriately constrained in the event that the number of dimensions p exceeds the number of individuals. Trouble arises, for example, if one desires to form gene clusters, where the count of genes per cluster can exceed the number of subjects. In the present discussion, we move forward based on the understanding that the appropriate sample size considerations are taken into account in estimation of Σ_c. Consider constraining the space of possible (feasible) models through Σ, by the decomposition of Σ_c into

$$\Sigma_k = \lambda_k E_k A_k E_k^T, \tag{7.20}$$

where E_k are orthogonal eigenvectors, A_k is a diagonal matrix with elements proportional to the eigenvalues, and λ_k is a constant of proportionality, for cluster k.

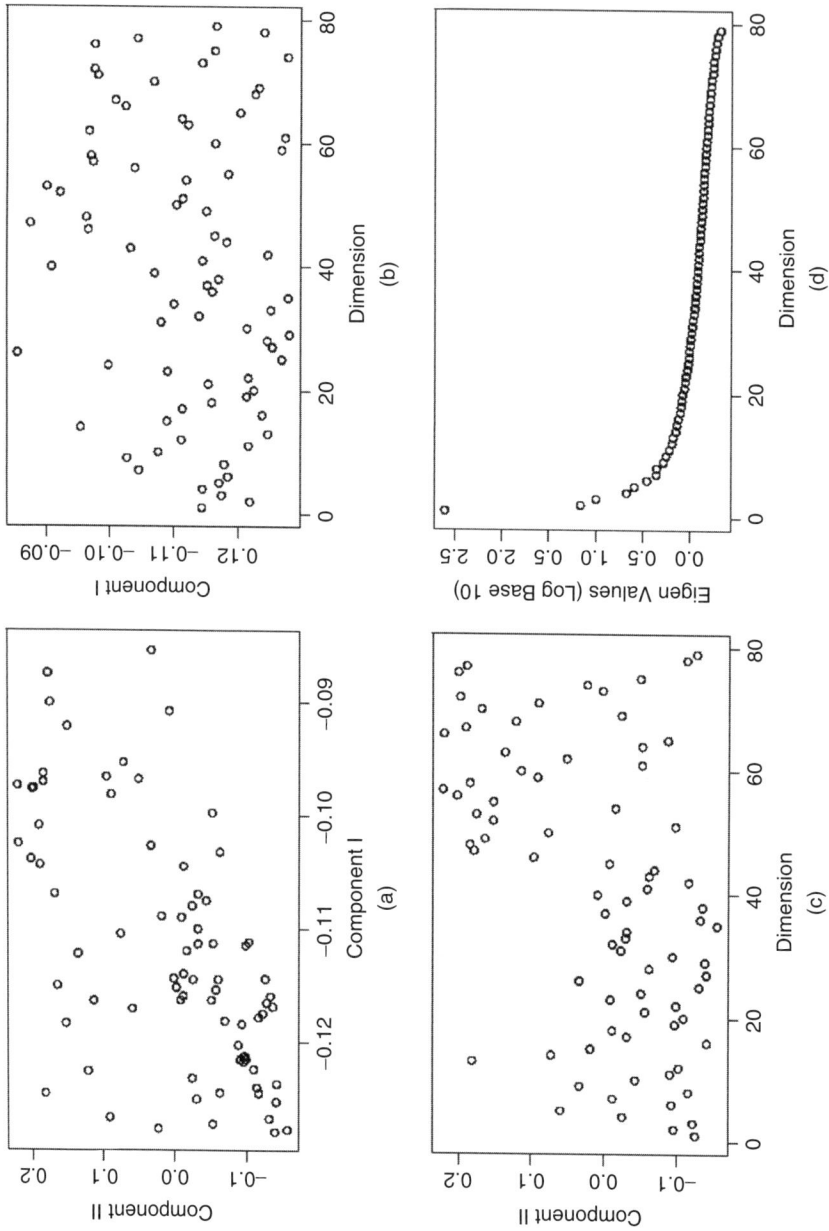

Figure 7.3   PC analysis: (a) scatter plot of first and second eigenvectors; (b) eigenvector 1; (c) eigenvector 2; (d) scree plot.

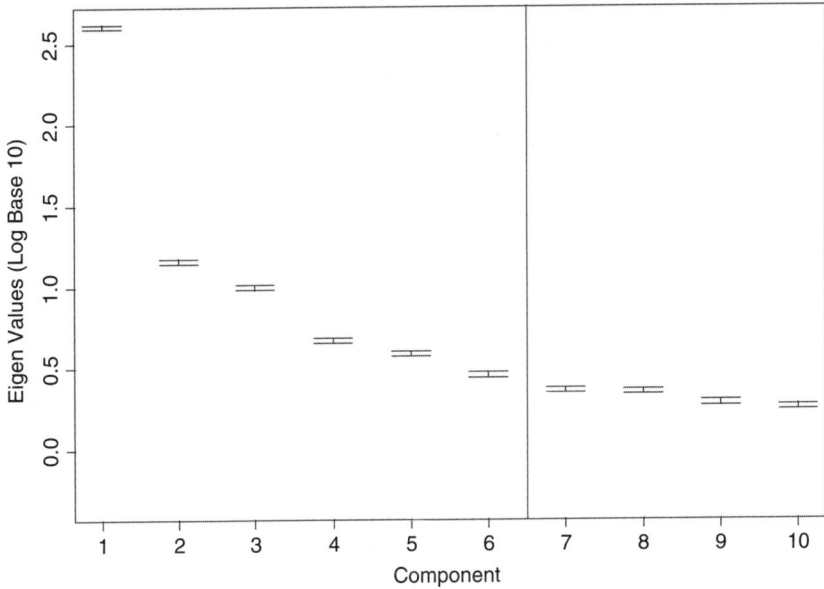

Figure 7.4   Posterior scree plot, with 95% posterior credible envelope.

Some specifications for Σ_k include the following:

Model	Specification
1	$\Sigma_k = \lambda I$
2	$\Sigma_k = \lambda_k I$
3	$\Sigma_k = \lambda_k A_k$
4	$\Sigma_k = \lambda_k E A_k E^T$
5	$\Sigma_k = \lambda_k E_k A_k E_k^T$

Model 1 treats restricts each cluster to an identical diagonal covariance matrix, with variance λ. The least restrictive, Model 5, is the most highly parameterized, allowing each cluster to have its rotation and scale. For each model, the EM algorithm is employed to maximize the likelihood. Convergence of the EM algorithm may be speeded up by considering good starting values. A sensible place to start is with the results form model-based hierarchical clustering.

Define models $m_1, m_2, m_3, \ldots$, each with a different specification on Σ_k, for $K = 1, 2, \ldots$ the total number of clusters. A strategy for model selection, based on the Bayesian information criterion (BIC), is as follows:

1. Determine the maximum number of cluster K to consider, and fix K.

2. Obtain starting values for each model specification and number of clusters $k = 1, \ldots, K$ from model based agglomerative hierarchical clustering, using the multivariate normal densities.

3. Apply the EM algorithm to fit each mixture model.

4. Compute the BIC for each mixture model, with the optimal EM fitted parameters.

This is the strategy proposed by Fraley and Raftery (2002).

The *Bayesian mixture model* is completed by specifying prior distributions for all unknown parameters, including model choice and the mixture weights $\pi_1, \ldots, \pi_K$. The conditional conjugate prior for $\pi_1, \ldots, \pi_K$ is the Dirichlet distribution,

$$P(\pi_1, \ldots, \pi_K | \alpha_1, \ldots, \alpha_K) = \frac{\Gamma\left(\sum_{k=1}^{K} \alpha_k\right)}{\prod_{k=1}^{K} \Gamma(\alpha_k)} \tau_1^{\alpha_1 - 1} \tau_2^{\alpha_2 - 1} \ldots \tau_K^{\alpha_K - 1}, \qquad (7.21)$$

with mean and variance

$$E(\pi_k) = \frac{\alpha_k}{\sum_l \alpha_l},$$

$$\mathrm{var}(\pi_k) = \frac{\alpha_k(\sum_{l \neq k} \alpha_l)}{(\sum_l \alpha_l)^2(\sum_l \alpha_l - 1)}. \qquad (7.22)$$

Since there is model uncertainty, i.e. as to the specification of Σ_k, the prior for model choice $\mathcal{M}$ is denoted $P(\mathcal{M} = m)$. The full model is:

$$\mathbf{x}_i | \mathcal{M}, \boldsymbol{\pi}, \theta \; i.i.d. \; \sum_{k=1}^{K(m)} \pi_k^{(m)} N(\mathbf{x}_i | \boldsymbol{\mu}_k^{(m)}, \Sigma_k^{(m)}),$$

$$\pi_k^{(m)} | \mathcal{M} \; \sim \; \mathrm{Dir}(\alpha, K(m)),$$

$$\boldsymbol{\mu}_k^{(m)}, \Sigma_k^{(m)} | \mathcal{M} \; \sim \; NIG(\boldsymbol{\mu}_k^{(m)}, \Sigma_k^{(m)}),$$

$$\mathcal{M} \; \sim \; P(\mathcal{M}). \qquad (7.23)$$

A convenient alternative specification introduces latent variables, h_{ik}, that indicate cluster membership, $h_{ik} = 1$ identifies subject i as a member of cluster k, and 0 otherwise. The h_{ik} are constrained so that $\Sigma_k h_{ik} = 1$ for all i. With this prior specification, the full model is:

$$\mathbf{x}_i \,|\, \boldsymbol{\pi}, \boldsymbol{\theta} \sim \sum_{k=1}^{K} P_k(\mathbf{x}_i | \boldsymbol{\theta}_k)^{h_{ik}},$$

$$h_{ik} \sim \text{Multinomial}(1, \boldsymbol{\pi}),$$

$$\pi_k \sim \text{Dir}(\alpha, R),$$

$$\theta_c k \sim P(\theta_k),$$

$$\mathcal{M} \sim P(\mathcal{M}). \tag{7.24}$$

The full conditionals for Gibbs sampling at iteration t, given $\mathcal{M} = m$, are:

1. $P(h_{ik}^{(t)} = 1 | \cdot) \propto \pi_k^{(t-1)} P_k(y_i | \theta_k^{(t-1)})$;

2. $P(\pi_k^{(t)} | \cdot) = \text{Dir}(\alpha_1 + \hat{n}_1^{(t)}, \ldots, \alpha_K + \hat{n}_K^{(t)})$, where $\hat{n}_k^{(t)} = \sum_i h_{ik}^{(t)}$; and

3. $P(\theta_k^{(t)} | \cdot) \propto P(\theta_k^{(t)}) \times \prod_i P_k(y_i | \theta_k^{(t)})^{\lambda_{ik}^{(t)}}$

One could fit all possible models, and compare them using, for example, the Bayes factor, although consider allowing model choice to be random, updated in the posterior sampling scheme by the reversible jump algorithm (Richardson and Green, 1997). The general idea for reversible jump, given current dimension model $\mathcal{M}$, is as follows:

1. From a staring state $(m, \boldsymbol{\theta}_m)$, propose a new model with probability q_{m,m^*}.

2. If m^* has a different dimensionality than m, generate an augmenting random dom u from the proposal $q(u | m, m^*, \boldsymbol{\theta}_m)$, otherwise

3. determine the proposed models parameters $\boldsymbol{\theta}_{m^*} = f_{m,m^*}(\boldsymbol{\theta}_m, u)$.

4. Accept the new model with probability $\min(r, 1)$ where

$$r = \frac{P(X | \boldsymbol{\theta}_{m^*}) P(\boldsymbol{\theta}_{m^*}) \pi_{m^*} q_{m,m^*} q(u^* | m^*, m, \boldsymbol{\theta}_{m^*})}{P(X | \boldsymbol{\theta}_K) P(\boldsymbol{\theta}_K) \pi_K q_{K^*,K} P(u | K, K^*, \boldsymbol{\theta}_K)} \left| \frac{\nabla f_{K,K^*}(\boldsymbol{\theta}_K, u)}{\nabla(\boldsymbol{\theta}_K, u)} \right|. \tag{7.25}$$

In the special case of the normal finite mixture model, at each step consider:

1. splitting one component into two, or combining two into one;

2. the birth or death of an empty component.

For (1) make a random choice between splitting and combining, according to predefined proposal probabilities, depending, of course, on K. If combined, choose two adjacent components, (k_1, k_2), at random based on the distance of their means. The merge is completed by setting

$$\pi_{k^*} = \pi_{k_1} + \pi_{k_2},$$

$$\pi_{k^*}\mu_{k^*} = \pi_{k_1}\mu_{k_1} + \pi_{k_2}\mu_{k_2},$$

$$\pi_{k^*}(\mu_{k^*}^2 + \sigma_{k^*}^2) = \pi_{k_1}(\mu_{k_1}^2 + \sigma_{k_1}^2) + \pi_{k_1}(\mu_{k_1}^2 + \sigma_{k_1}^2), \qquad (7.26)$$

matching first and second moments. The reverse split begins by choosing a group k^* at random and generating $u_h \sim \text{Beta}(2, 2)$, $h = 1, 2, 3$:

$$\pi_{k_1} = \pi_{k^*}u_1, \qquad (7.27)$$

$$\pi_{k_2} = \pi_{k^*}(1 - u_1), \qquad (7.28)$$

$$\mu_{k_1} = \mu_{k^*} - u_2\sigma_{k^*}\sqrt{\pi_{k_1}/\pi_{k_2}}, \qquad (7.29)$$

$$\mu_{k_2} = \mu_{k^*} + u_2\sigma_{k^*}\sqrt{\pi_{k_2}/\pi_{k_1}}, \qquad (7.30)$$

$$\sigma_{k_1}^2 = u_3(1 - u_2^2)\sigma_{k^*}\pi_{k^*}/\pi_{k_1}, \qquad (7.31)$$

$$\sigma_{k_2}^2 = (1 - u_3)(1 - u_2^2)\sigma_{k^*}\pi_{k^*}/\pi_{k_2}. \qquad (7.32)$$

For a birth or death the same proposal probabilities are used. If the draw is a birth, the parameters are drawn from

$$\pi_{k^*} \sim \text{Beta}(1, \pi_0), \qquad (7.33)$$

$$\mu_{k^*} \sim N(\xi, \kappa^{-1}), \qquad (7.34)$$

$$\sigma_{k^*}^{-2} \sim \text{Gamma}(a, b), \qquad (7.35)$$

and the existing weights are rescaled. For a death, a component is deleted, and the existing weights are rescaled.

7.8.1   Label Switching

Mixture type models in general suffer from a lack of cluster label identifiability, as the cluster labels are not necessarily ordered without further constraints, and as such are purely semantic. Label switching occurs when, during the course of posterior updating, the labels of respective components are switched. This may not seem like a crucial problem, at least as long as the component integrity is maintained, although it can pose serious challenges for interpreting the results of an MCMC algorithm. Since the labels are arbitrary, note that the likelihood

$$X|\mathcal{P}, \theta \sim \prod_i \sum_k \pi_{\mathcal{P}(k)} P(\mathbf{x}_i; \theta_{\mathcal{P}(k)}) \qquad (7.36)$$

is invariant to any random permutation $\mathcal{P}(k)$ of $k = 1, \ldots, K$. Some authors argue that MCMC convergence is not achieved until all possible label permutations have been realized on one's Monte Carlo chains. Efficient algorithms have been proposed. Some possible solutions are based on:(1) relabeling algorithms, (2) identifiability constraints, (3) strong priors, (4) integrating out θ_k, and (5) label invariant loss functions. One proposal for relabeling is to randomly permute the labels (Frühwirth-Schnatter, 2001). Strong priors can also be helpful, or for example imposing an order constraint $\mu_1 < \mu_2 < \ldots$ on the means. For more on the label switching problem, see Jasra et al. (2005).

Example 7.3

We analyzed the yeast galactose data of Ideker et al. (2001) where four replicate hybridizations were performed for each cDNA array experiment. We used a subset of 205 genes that are reproducibly measured, whose expression patterns reflect four functional categories in the GO listings and that we expect to cluster together. For these data, our goal is to cluster the genes using mclust and the four functional categories are used as our external knowledge.

Figure 7.5 shows the available options in mclust for hierarchical clustering (HC) and expectation maximization (EM). The model identifiers code geometric characteristics of the model. For example, EFV denotes a model in which volumes of all clusters are equal (E), shapes of all clusters are fixed (F), and the orientation of the clusters are allowed to vary (V). Parameters associated with characteristics designated by either E or V are determined from the data.

ID	Model	HC	EM	Distribution	Volume	Shape	Orientation
EI	λI	×	×	Spherical	equal	equal	NA
VI	$\lambda_k I$	×	×	Spherical	variable	equal	NA
EEE	$\lambda D A D^T$	×	×	Ellipsoidal	equal	equal	equal
VVV	$\lambda_k D_k A_k D_k^T$	×	×	Ellipsoidal	variable	variable	variable
EFV[2]	$\lambda D_k \hat{A} D_k^T$		×	Ellipsoidal	equal	fixed	variable
EEV	$\lambda D_k A D_k^T$		×	Ellipsoidal	equal	equal	variable
VFV[2]	$\lambda D_k \hat{A} D_k^T$	×		Ellipsoidal	variable	fixed	variable
VEV	$\lambda_k D_k A D_k^T$		×	Ellipsoidal	variable	equal	variable

Figure 7.5   Model descriptions in mclust.

We can see from the BIC in Figure 7.6 that BIC is maximized for the VEV model at two parameters. So mclust is saying that there are two clusters in the data. VEV denotes a model in which volumes of all clusters vary, shapes of all clusters are equal, and the orientation of the clusters are allowed to vary. The two clusters can be seen in Figure 7.7 when the data are projected into two dimensions. The red squares are one cluster and the blue triangles the other. The ellipses show the orientation of the cluster in two dimensions.

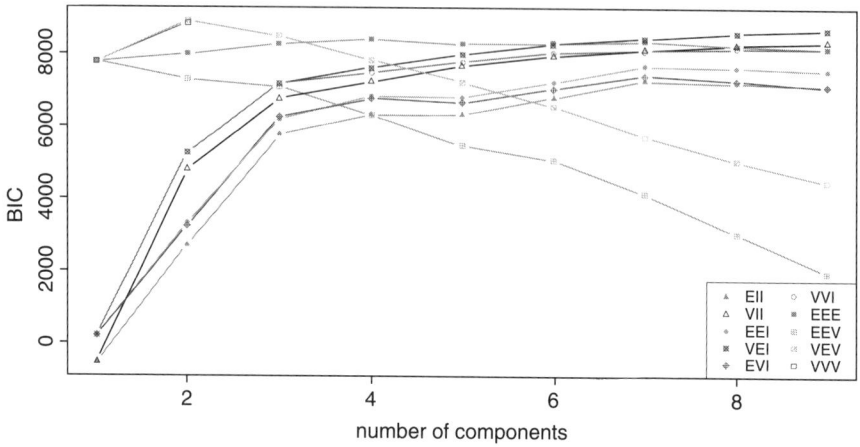

Figure 7.6   BIC values for each of the possible models in `mclust`.

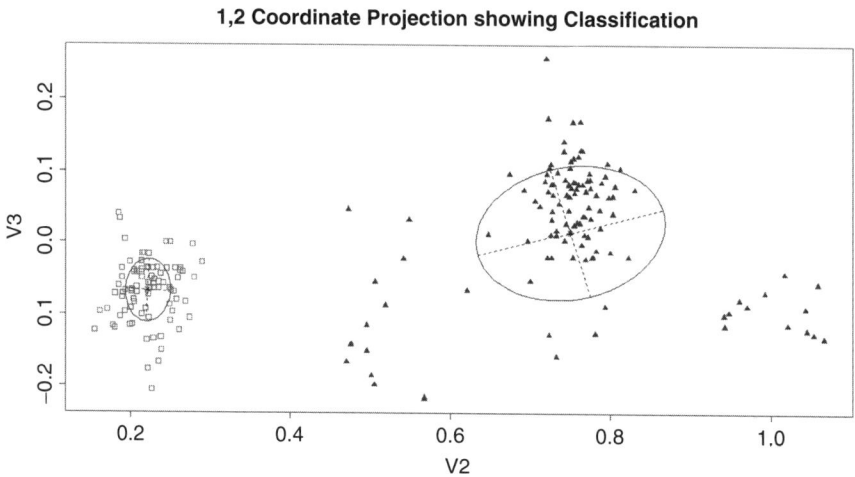

Figure 7.7   Clustering profile in `mclust`.

7.9   Clustering Using Dirichlet Process Prior

The alternative Bayesian clustering method is the use of infinite mixture model via the Dirichlet process prior (Ferguson, 1973; Antoniak, 1974; Neal, 2000). In contrast to the finite mixture approach, this model does not require the number of mixture components to be specified. Following our notation, x_i is the multivariate gene expression profile for the ith tissue sample, which is assumed to be drawn from a distribution $P(\theta)$ (where θ could be multidimensional as in Section 7.8), with the prior (mixing) distribution over θ being G. In the usual parametric Bayesian modeling setup, we assign some parametric distribution for G and use further prior distributions for those unknown hyperparameters corresponding to that parametric distribution. In a nonparametric problem we assume that G is completely unknown and assign stochastic process based prior distributions over the class of distribution functions. The most popular nonparametric priors are Dirichlet processes. They were introduced by Ferguson (1973) with additional clarification provided by Blackwell and MacQueen (1973), Blackwell (1973), Antoniak (1974), Sethuraman and Tiwari (1982) and Sethuraman (1994).

A cumulative distribution function or, equivalently, a probability measure G on Θ is said to follow Dirichlet process, i.e. $G \sim DP(\alpha G_0)$, if, for any measurable partition of Θ, $B_1, B_2, \ldots, B_m$, the joint distribution of the random variable $[G(B_1), G(B_2), \ldots, G(B_m)$ follows a Dirichlet distribution $\text{Dir}(\alpha G_0(B_1), \ldots, \alpha G_0(B_m))$. Here G_0 is a specified probability measure interpreted as the prior mean or baseline distribution of G or $E(G) = G_0$. α is the precision parameter showing the a priori strength of belief in G_0, so the larger α is the closer we expect G to be to G_0 a priori.

Our main aim in this section is to use Dirichlet processes as a basis for model-based clustering. An early criticism which has been leveled at Dirichlet process models is that they assign mass 1 to the subset of all discrete distribution on T. However, currently this inherent discreteness property has been utilized successfully for clustering. In particular, if we generate two independent sequences of i.i.d. random variables, θ_i and z_i, such that $\theta_i \sim G_0$ and $z_i \sim \text{Beta}(1, \alpha)$ and let $w_i = z_i \prod_{j<i}(1 - z_j)$. Then the random function $G = \sum_{i=1}^{\infty} w_i \delta(\theta_i)$ where $\delta(\theta_i)$ is a degenerate distribution at θ_i and it is clear that $\sum w_i = 1$. This characterizes $DP(\alpha G_0)$ as $G \sim DP(\alpha G_0)$ if and only if almost every realization is of this form. This is a surely discrete distribution hence the realizations from this distribution will have positive probability of a tie which is the basis of the cluster formation. Furthermore, the realizations w_i determine an infinite breaking of a stick of unit length. We first break off a piece of length w_1, then a piece of length w_2, etc., hence it is also known as a stick breaking prior (Freedman, 1963).

The conjugacy result states that given a set of n realizations $\boldsymbol{\theta} = (\theta_1, \ldots, \theta_n)$ from $G \sim DP(\alpha G_0)$, the posterior distribution of G given $\boldsymbol{\theta}$ is also a Dirichlet process. More precisely, $G|\boldsymbol{\theta} \sim DP(\alpha^*, G_0^*)$ where $\alpha^* = \alpha + n$ and $G_0^* = (\alpha + n)^{-1}[\alpha G_0 + \sum \delta(\theta_i)]$. Though $\theta_1, \ldots, \theta_n$ are independent draws from G, if we marginalize over G then the θ_i are no longer independent since they shared a common random G. This marginal distribution is built sequentially using appropriate G_0^*s. The sequential scheme is, we draw θ_1 from G_0, we draw $\theta_2|\theta_1$ from the distribution $(\alpha + 1)^{-1}[\alpha G_0 + \delta(\theta_1)]$, $\ldots$, up to $\theta_n|\theta_1, \ldots, \theta_{n-1} \sim (\alpha + (n - 1))^{-1}[\alpha G_0 + \sum \delta(\theta_i)]$.

In general, we can write the conditional distribution for the ith θ as

$$\theta_i | \theta_1, \ldots, \theta_{i-1} \sim \frac{1}{i-1+\alpha} \sum_{j=1}^{i-1} \delta(\theta_j) + \frac{\alpha}{i-1+\alpha} G_0. \qquad (7.37)$$

It is clear that this conditional distribution is a mixture of a continuous distribution and a point mass. This conditional distribution relates the Dirichlet process prior with clustering idea. It says that a priori for the ith realization there is a probability of $(i-1+\alpha)^{-1}$ that it will be equal to (or cluster with) any of the previous $i-1$ realizations, $\theta_1, \ldots, \theta_{i-1}$, whereas the probability of a new draw from the baseline distribution G_0 is $\alpha/(i-1+\alpha)$.

A convenient way to obtain these results is to consider the equivalent finite mixture models as in Section 7.8 and taking the limit as K goes to infinity (Neal, 2000). The hierarchical model we shall consider is

$$x_i | c_i, \boldsymbol{\theta} \sim P(\theta_{c_i}),$$

$$c_i | \mathbf{p} \sim Discrete(p_1, \ldots, p_K), \quad c_i \in \{1, \ldots, K\},$$

$$\theta_c \sim G_0,$$

$$\mathbf{p} \sim \text{Dir}(\alpha/K, \ldots, \alpha/K), \qquad (7.38)$$

where c_i indicates the cluster membership of x_i and θ_c indicates the cluster-specific parameter corresponding to the cth group. The mixing proportions for the classes, $\mathbf{p} = (p_1, \ldots, p_K)$ have been assigned a Dirichlet prior, with concentration parameter α/K which approaches 0 as K goes to infinity.

By integrating out $\mathbf{p}$ we can write the prior for the c_i as the product of the conditional probabilities as shown below:

$$
\begin{aligned}
P(c_i = c | c_1, \ldots, c_{i-1}) &= \frac{P(c_1, \ldots, c_{i-1}, c_i = c)}{P(c_1, \ldots, c_{i-1})} \\
&= \frac{\int_{\mathbf{p}} P(c_1, \ldots, c_{i-1}, c | \mathbf{p}) f(\mathbf{p}) d\mathbf{p}}{\int_{\mathbf{p}} P(c_1, \ldots, c_{i-1} | \mathbf{p}) f(\mathbf{p}) d\mathbf{p}} \\
&= \frac{\int_{\mathbf{p}} P(c_1|\mathbf{p}) \ldots P(c_{i-1}|\mathbf{p}) P(c|\mathbf{p}) \prod_{i=1}^{K} p_i^{\frac{\alpha}{K}-1} d\mathbf{p}}{\int_{\mathbf{p}} P(c_1|\mathbf{p}) \ldots P(c_{i-1}|\mathbf{p}) \prod_{i=1}^{K} p_i^{\frac{\alpha}{K}-1} d\mathbf{p}} \\
&= \frac{\int_{\mathbf{p}} p_1^{n_{i,1}+\frac{\alpha}{K}-1} \ldots p_c^{n_{i,c}+\frac{\alpha}{K}-1+1} \ldots p_K^{n_{i,K}+\frac{\alpha}{K}-1} d\mathbf{p}}{\int_{\mathbf{p}} p_1^{n_{i,1}+\frac{\alpha}{K}-1} \ldots p_c^{n_{i,c}+\frac{\alpha}{K}-1} \ldots p_K^{n_{i,K}+\frac{\alpha}{K}-1} d\mathbf{p}}
\end{aligned}
$$

where $n_{i,l}$ is the number of c_l for $l < i$ that are equal to $l \in \{0, K\}$. The above integration yields the normalizing constant of a Dirichlet distribution, hence we obtain

$$P(c_i = c|c_1, \ldots, c_{i-1}) = \frac{\frac{\Gamma(n_{i,1}+\frac{\alpha}{K})\ldots\Gamma(n_{i,c}+\frac{\alpha}{K}+1)\Gamma(n_{i,K}+\frac{\alpha}{K})}{\Gamma(i+\alpha)}}{\frac{\Gamma(n_{i,1}+\frac{\alpha}{K})\ldots\Gamma(n_{i,c}+\frac{\alpha}{K})\Gamma(n_{i,K}+\frac{\alpha}{K})}{\Gamma(i-1+\alpha)}}$$

$$= \frac{\Gamma(n_{i,c} + \frac{\alpha}{K} + 1)\Gamma(i - 1 + \alpha)}{\Gamma(n_{i,c} + \frac{\alpha}{K})\Gamma(i + \alpha)}$$

$$= \frac{n_{i,c} + \frac{\alpha}{K}}{i - 1 + \alpha}. \tag{7.39}$$

Now letting $K \to \infty$, the prior of c_i is

$$P(c_i = c|c_1, \ldots, c_{i-1}) \to \frac{n_{i,c}}{i - 1 + \alpha}$$

$$P(c_i \neq c_j|c_1, \ldots, c_{i-1}) \to 1 - \sum_{c=1}^{\overset{K\to\infty}{}} \frac{n_c}{i - 1 + \alpha} \to \frac{\alpha}{i - 1 + \alpha} \tag{7.40}$$

Note: Even though $P(c_i \neq c_j|c_1, \ldots, c_{i-1}) = 0$ for finite K it is > 0 when $K \to \infty$.

It is clear that the limit of this model as C goes to infinity is equivalent to the Dirichlet process mixture model due to the correspondence between the conditional probability of θ_i in equation (7.37) and equation (7.40). Moreover, we can represent this joint prior distribution of θ as product of conditional distributions after marginalizing G. We combine this prior with the data to make posterior inferences about the clustering process.

A colorful description of this clustering procedure and its corresponding cluster indicator distribution is given by the *Chinese restaurant process* presented in Arratia et al. (1992). Here, customers enter a restaurant sequentially and sit one after another. Initially, all tables are folded up, so that one table is opened when the first customer enters. After customers $1, \ldots, i - 1$ are seated, the ith customer will either choose an empty table with probability $\alpha/(i - 1 + \alpha)$ for some $\alpha > 0$ or an occupied table with probability proportional to the number of occupants at the given table. This scheme is same as the representation in equation (7.40).

Posterior Inference The hierarchical model in this setup is

$$x_i|\theta_i \sim P(\theta_i),$$

$$\theta_i|G \sim G, \tag{7.41}$$

$$G \sim DP(G_0, \alpha).$$

Now assigning a Dirichlet process prior for G, a priori the full conditional distribution (conditioned on all the θs) for θ_i will be

$$\theta_i | \theta_{-i} \sim \frac{1}{n-1+\alpha} \sum_{j \neq i} \delta(\theta_j) + \frac{\alpha}{n-1+\alpha} G_0. \qquad (7.42)$$

This conditional prior can be derived from equation (7.37) by imagining that i is the last of the n observations, as we may, since the observations are exchangeable so the ordering does not matter. When combined with the likelihood, the conditional distribution we obtain is

$$\theta_i | \theta_{-i}, x_i \sim \sum_{j \neq i} q_{i,j} \delta(\theta_j) + r_i H_i. \qquad (7.43)$$

Here H_i is the posterior distribution for θ based on the baseline prior G_0 and the single observation x_i, with likelihood $P(x_i | \theta)$. $q_{i,j}$ and r_i are the mixing weights whose values are defined as $q_{i,j} = b P(x_i, \theta_j)$ and $r_i = b\alpha \int P(x_i | \theta) dG_0(\theta)$ where b is such that $\sum_{j \neq i} q_{i,j} + r_i = 1$. We only consider the conjugate case where the baseline prior G_0 is conjugate to the likelihood F so that computing the integral defining r_i and sampling from H_i are feasible operations. In this conjugate case we can use equation (7.43) as the conditional distribution in our Gibbs sampling framework (Escobar 1994; Escobar and West, 1995). To obtain a more suitable algorithm we can draw the cluster indicators directly as has been shown in prior modeling in equation (7.40). The argument is similar to the prior modeling situation where we consider a finite mixture model as in (7.38), then integrate out the mixing proportion p. When K is finite, within each Gibbs iteration, a new value will be picked for each clustering indicator c_i given the data x_i, cluster specific parameter θ_c, and all other c_j for $j \neq i$ (written as c_{-i}). Then we draw a new value for each θ_c conditioned on the the the data x_i and $c_i = c$. The conditional probabilities for c_i will be

$$P(c_i | c_{-i}, x_i, \theta) = b P(x_i | \theta_c) \frac{n_{-i,c} + \alpha/K}{n-1+\alpha}, \qquad (7.44)$$

where $n_{-i,c}$ is the number of c_j for $j \neq i$ that are equal to c, and b is the appropriate normalizing constant. The calculation of this probability is very similar to equation (7.39) except that the conditional prior (as in equation (7.42)) has been multiplied by the likelihood $P(x_i | \theta_c)$ and except for using the exchangeability argument so that we can consider i as the last observation. That way we calculate the full conditional probability rather than sequential conditional probability as in equation (7.40).

When C goes to infinity, we consider only those θ_c that are currently associated with some observations. The conditional distributions for the cluster indicators will be

$$P(c_i = c|c_i, x_i, \boldsymbol{\theta}) = b\frac{n_{-i,c}}{n-1+\alpha}P(x_i|\theta_c),$$

$$P(c_i \neq c_j \forall j \neq i|c_i, x_i, \boldsymbol{\theta}) = b\frac{\alpha}{n-1+\alpha}\int P(x_i|\theta)dG_0(\theta), \qquad (7.45)$$

where $\boldsymbol{\theta}$ is the set of θ_c currently associated with one of the clusters and b is the normalizing constant that makes the above probability sum to one.

The numerical values of the c_i are arbitrary and they only represent whether or not $c_i = c_j$. They determine the configuration in which the data items are grouped in accordance with shared values for θ. The numerical values of c_i are usually chosen for programming convenience. In a Gibbs iteration, if c_i chooses a value not equal to any other c_j, a value for θ_{c_i} is chosen from H_i, the posterior distribution based on the prior G_0 and the observation x_i.

The Gibbs sampling algorithm works as follows (Bush and MacEachern, 1996; West et al.,1994). In each iteration of the Gibbs sampling we conditionally draw from the states $\mathbf{c} = (c_1, \dots, c_n)$ and $\boldsymbol{\theta} = (\theta_c : c \in \{c_1, \dots, c_n\})$:

(i) For $i = 1, \dots, n$, if the present value of c_i is not associated with any other observation ($n_{-i,c_i} = 0$) then remove the corresponding θ_{c_i} from the state. Draw a new value for c_i from $c_i|c_{-i}, x_i, \boldsymbol{\theta}$ using equation (7.45). If the new c_i is not associated with any other observation, draw a value for θ_{c_i} from H_i and add it to the state.

(ii) For all $c \in \{c_1, \dots, c_n\}$, draw a new value from θ_c conditioned on all the data X for which $c_i = c$. that is, from the posterior distribution based on the prior G_0 and all the data points currently associated with latent class c.

This algorithm can be made more efficient by further marginalization of the parameters θ in a conjugate setup (MacEachern, 1994; Neal, 1992). In this case the c_is are updated using the conditional probabilities:

$$P(c_i = c|c_i, x_i) = b\frac{n_{-i,c}}{n-1+\alpha}\int P(x_i|\theta)dH_{-i,c}(\theta),$$

$$P(c_i \neq c_j \forall j \neq i|c_i, x_i, \boldsymbol{\theta}) = b\frac{\alpha}{n-1+\alpha}\int P(x_i|\theta)dG_0(\theta), \qquad (7.46)$$

where $H_{-i,c}$ is the posterior distribution of ϕ based on the prior G_0 and all the observations x_j for which $j \neq i$ and $c_j = c$. In this situation the Gibbs sampling contains one step to draw c_i from $c_i|c_{-i}, x_i$ as defined in equation (7.46).

7.9.1 Infinite Mixture of Gaussian Distributions

Medvedovic and Sivaganesan (2002) used an infinite mixture of Gaussian distributions to cluster gene expression data using the methodology described in the previous section. They use the Gaussian likelihood for $P(x|\theta)$ as well as an assigned Gaussian distribution for the baseline prior G_0. Hence the hierarchical model in (7.41) adopted to

$$x_i | \theta_i \sim N(\mathbf{\mu}_i, \sigma_i^2 I),$$

$$\mathbf{\mu}_i, \sigma_i^2 | G \sim G,$$

$$G \sim DP(G_0, \alpha), \qquad (7.47)$$

where G_0 is NIG distributed. Hence $\mathbf{\mu}_i | \lambda, r \sim N(\lambda, r^{-1} I)$ and $\sigma_i^2 \sim$ Gamma$(a/2, ab/2)$. They have further hierarchical structure given by

$$\lambda | \mathbf{\mu}_x, \sigma_x^2 \sim N(\mathbf{\mu}_x, \sigma_x^2 I),$$

$$a \sim \text{Gamma}(0.5, 0.5),$$

$$b | \sigma_x^2 \sim \text{Gamma}\left(0.5, \frac{1}{2\sigma_x^2}\right),$$

$$G \sim DP(G_0, \alpha),$$

where $\mathbf{\mu}_x$ is the average of all profiles analyzed and σ_x^2 is the average of gene-specific sample variances based on all profiles analyzed. They assigned the precision parameter $\alpha = 1$ though further priors could be assigned on α. This is not exactly a conjugate setup but the corresponding Gibbs sampling algorithm had been described in that paper. The conditional distribution of the cluster can be obtained easily using equation (7.45) with specific Gaussian likelihood and the above mentioned prior structure.

We can use the conditional distributions in a Gibbs sampling framework to generate clusters from the posterior distribution. To identify the best cluster (or some good ones) out of these samples itself is a challenging problem. The maximum a posteriori (MAP) approach usually fails in this situation as the state space of possible clustering combinations can be very large and different clustering occurs frequently with very small posterior probabilities. Quintana and Iglesias (2003) provide a search algorithm to approach the best model by minimizing a penalized risk, but for large data sets the computational constraints can be prohibitive. Medvedovic and Sivaganesan used the sequence of clustering generated by the Gibbs sampler after the burn-in cycles to calculate pairwise probabilities for two genes to be generated by the same pattern. This pairwise probability is estimated by the relative frequency with which these two genes are in the same cluster among the MCMC samples. This probability measure can be treated as a distance measure to calculate pairwise distances between the genes. Based on these distances, clusters of similar expression profiles are created by the complete linkage approach (Everitt, 1993).

Example 7.4

We have used the Gaussian infinite mixture model (GIMM) code provided by Medavedovic to reanalyze the yeast galactose data using 205 genes. To run this

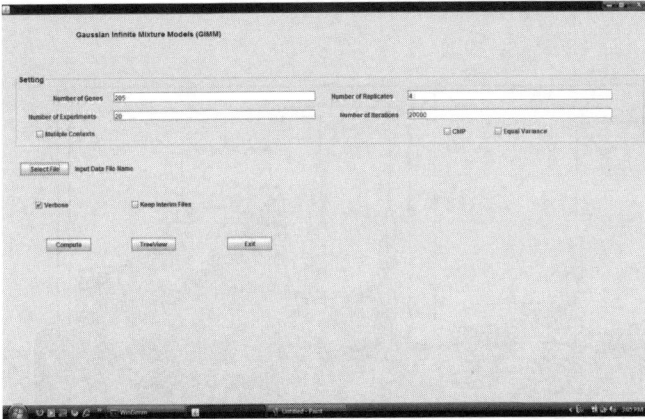

Figure 7.8   Starting up WINGIMM.

code, the first step is to start the GIMM program (see Figure 7.8). Then input all the required fields and press **Compute**. The program will compute the full clustering configuration. After completion press **TreeView** and you will get a tree view window from which you can interpret the clustering configuration of the genes as in Figure 7.9. The final clustering configuration for the data is shown in Figure 7.10.

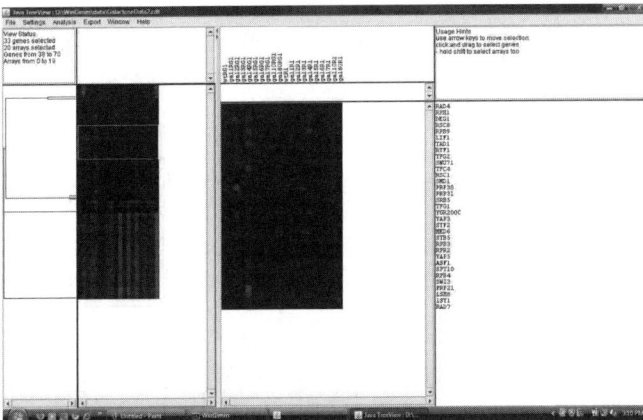

Figure 7.9   Tree in WINGIMM.

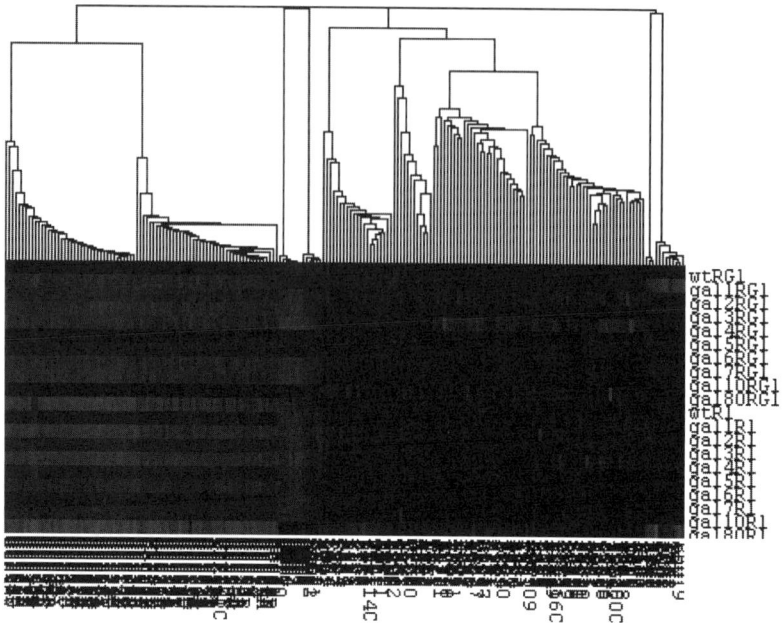

Figure 7.10   Clustering tree in GIMM.

We identify the number of clusters using the adjusted Rand index (ARI: Hubert and Arabie, 1985) which is derived from the Rand index (Rand, 1971). Suppose n objects have been classified by two separate partitions A and B. Let n_{ij} be the number of elements classified by both $A_i \in A$ and $B_j \in B$. The Rand index is evaluated as

$$ARI = \frac{\text{Data} - \text{Expected index}}{\text{Maximum value} - \text{Expected index}},$$

which is a number bounded between 0 and 1, 1 meaning perfect correspondence. Assuming a generalized hypergeometric distribution, the expected value for the index is

$$E\left[\sum_{i,j} \binom{n_{ij}}{2}\right] = \frac{\left[\sum_i \binom{n_{i.}}{2} \sum_j \binom{n_{.j}}{2}\right]}{\binom{n}{2}}.$$

The corresponding adjusted Rand index (ARI) is then derived (Hubert and Arabie, 1985) as

$$
ARI = \frac{\sum_{i,j}\binom{n_{ij}}{2} - \left[\sum_i\binom{n_{i.}}{2}\sum_j\binom{n_{.j}}{2}\right]/\binom{n}{2}}{\frac{1}{2}\left[\sum_i\binom{n_{i.}}{2}+\sum_j\binom{n_{.j}}{2}\right] - \left[\sum_i\binom{n_{i.}}{2}\sum_j\binom{n_{.j}}{2}\right]/\binom{n}{2}}.
$$

(7.48)

The expression patterns of these genes reflect four functional categories in the GO Consortium (Ashburner et al., 2000), hence these classes have been used as the external criterion to assess clustering results. We tried several different cluster sizes and patterns in GIMM and calculated corresponding ARIs. The best result was obtained with an ARI of 0.95. The `mclust` package, which uses the BIC to determine the number of clusters to be 2, produces an ARI value of 0.78. We observe that genes in the same GO category may not be co-expressed and we need to be careful to select the external criterion for assessing clustering results. More formal Bayesian model choice procedures should be utilized to compare clustering results (Ray and Mallick, 2006).

8

Bayesian Graphical Models

8.1 Introduction

In Chapter 7 we used cluster analysis to identify co-regulated genes. Recently there has been growing interest in the underlying relationships between genes. For example, we may wish to obtain a network of dependencies between the differentially expressed genes. In this chapter we do further investigation to develop probability models to relate genes based on microarray data. In this way we can extract more information from the data to develop biologically meaningful relationships between genes. We describe Bayesian graphical models to represent these biological processes and we make posterior inference about these models based on available gene expression data.

Recently several methods have been developed for modeling regulatory and cellular networks based on genome-wide high-throughout data, including both Bayesian network modeling (Friedman, 2004; Segal et al., 2003) and Gaussian graphical modeling (Schaffer and Strimmer, 2005; Dobra et al., 2004). The goal of such probability models is to investigate the patterns of association in order to generate biological insights plausibly related to underlying biological and regulatory pathways. The interaction between two genes in a gene network defined by such graphical models does not necessarily imply a physical interaction, but rather may refer to an indirect regulation via proteins, metabolites and noncoding RNA (Bansal et al., 2007).

Bayesian Analysis of Gene Expression Data   B. Mallick, D. Gold, and V. Baladandayuthapani
© 2009 John Wiley & Sons, Ltd

8.2   Probabilistic Graphical Models

A central goal of all of our studies in this section is to construct a model for genetic networks such that the model class incorporates dependencies between genes. To understand how cells function, it is necessary to study the behavior of genes in a holistic rather than in an individual manner because the expressions and activities of genes are not isolated nor independent of each other.

Probabilistic graphical models are the basic tool to represent these dependence structures. They provide a simple way to visualize the structure of a probability model as well as providing insights into the properties of the model, including conditional independence structures. A graph compromises vertices (nodes) connected by edges (links or arcs). In a probabilistic graphical model, each vertex represents a random variable (scalar or vector) and the edges express probabilistic relationship between these variables. The graph defines the way in which the joint distribution over all the random variables can be decomposed into a product of factors contacting subset of the variables. There are two types of probabilistic graphical models: directed graphical models (DAGs) or Bayesian networks where the edges of the graphs have a particular directionality which expresses causal relationships between random variables; and (ii) undirected graphical models in which the edges do not carry the directional information. In this book we concentrate on Bayesian networks; however, Gaussian undirected graphical models are also promising tools to explore relationships between genes. For further information about undirected Gaussian graphical models, see Schaffer and Strimmer (2005), Dobra et al. (2004), Wei and Li (2007), and references therein. In addition, there have been a number of alternative ways to model gene regulatory networks, including neural networks (Weaver et al., 1999), differential equations (Mestl et al., 1995), and Boolean networks (Huang, 1999).

8.3   Bayesian Networks

Bayesian networks are graphical models that explicitly represent probability relationships between variables $x_1, x_2, \ldots, x_n$ (Pearl, 1988; Jensen, 1996). Suppose that x_i represents the mRNA expression level of a gene. x_i could be continuous for original expression data, or discrete when the expression levels are discretized into two or more categories (e.g. expressed/not expressed, high/medium/low), using appropriate thresholds. The model structure embeds conditional dependencies and independencies and efficiently specifies the joint probability distribution of all the variables. Bayesian networks describe joint processes processes composed of *locally* interacting components; that is, the value of each component directly depends on the value of a relatively small number of parent components. Through this Bayesian network we obtain the graphical model of joint probability distributions that capture properties of conditional independence between variables. This local nature of the Bayesian network is attractive for computational efficiency.

By way of illustration, we consider a situation with five genes A, B, C, D, E (8.1). Here x_i, $i = A, \ldots, E$, represents the gene expression corresponding to the ith gene and the joint distribution of xs is specified by a conditional distribution defined by the graphical model. Suppose that gene B is a transcription factor of gene C. Gene A does not directly affect C, and once we fix the expression level of B we will observe that A and C are independent. In other words, the effect of A on C is mediated through B. In this case we say that A and C are conditionally independent given B. Similarly, B and D are regulated by A. Thus, B and D are conditionally independent once we know the expression level of A. Gene E inhibits the transcription of A and in our graphical model we represent it by placing an arc form E to B. The expression of B is regulated by two genes A and E which are known as B's parents, denoted as $Pa(B)$. Hence the joint probability distribution can be specified based on the graphical structure as $P(x_A, x_B, x_C, x_D, x_E) = P(x_C|x_B)P(x_B|x_A, x_E)P(x_D|x_A)P(x_A)P(x_E)$. The procedure for writing the the joint distribution is based on sequential decomposition of the joint probability distribution through conditional distributions and use of conditional independence arguments based on the graph. For example, $P(x_A, x_B, x_C, x_D, x_E) = P(x_C|x_A, x_B, x_D, x_E)P(x_A, x_B, x_D, x_E)$. Now from the graph it is clear that x_C only depends on x_B and, conditional on x_B (i.e. if we know the value of x_B), it is independent of the other three random variables. Hence we can express $P(x_C|x_A, x_B, x_D, x_E)$ as $P(x_C|x_B)$. Next we can further decompose $P(x_A, x_B, x_D, x_E)$ as $P(x_B|x_A, x_D, x_E)P(x_A, x_D, x_E)$. In the same way, using the dependence structure of the graph and using the conditional independence argument, $P(x_B|x_A, x_D, x_E) = P(x_B|x_A, x_E)$. Similarly, decompose $P(x_A, x_D, x_E) = P(x_D|x_A, x_E)P(x_A)P(x_E)$. Due to conditional independence, $P(x_D|x_A, x_E) = P(x_D|x_A)$. Also x_A and x_E are marginally independent hence $P(x_A, x_E) = P(x_A)P(x_E)$. After decomposition of these conditional distributions and multiplying them properly we obtain the joint distribution. Hence the whole network model can presented in terms of the relationship between parents and children, and local models can be composed through conditional dependence assumptions.

In general, in our application the variables are expression levels of genes and the relationships between these variables are represented by a directed graph in which vertices correspond to variables and directed edges between vertices represent their dependencies. The specification contains two components. The first component consists of a directed acyclic graph $G(V, E)$ with a node set V corresponding to the random variables $x_1, \ldots, x_n$, and an edge set E on these nodes. The edges capture the conditional independence structure in the graphical model. The existence of an edge between two nodes implies conditional dependence between them. A node is conditionally independent of all other nodes given its parents in the networks. The second component is a set of conditional probability distributions for each node in the graph G. The probability distributions are locally specified by the conditional distribution $P\{x_i|Pa(x_i)\}$.

The overall list of marginal and conditional independencies represented by the directed acyclic graph is summarized by the local and global Markov properties. The local Markov property states that each node is independent of its nondescendants given the parent nodes and leads to a direct factorization of the joint distribution of the network variables into the product of the conditional distribution of each variable x_i given its parents $Pa(x_i)$. For example, in Figure 8.1 we have a network of five variables. As shown above, $P(x_A, x_B, x_C, x_D, x_E) = P(x_C|x_B)P(x_B|x_A, x_E)P(x_D|x_A)P(x_A)P(x_E)$, where we first replaced $P(x_C|x_A, x_B, x_D, x_E)$ by $P(x_C|x_B)$ due to the conditional independence assumption. Now x_B is the parent of x_C, hence $P(x_C|x_B) = P(x_C|Pa(x_C))$. A similar argument works for all other conditional distributions in the expression.

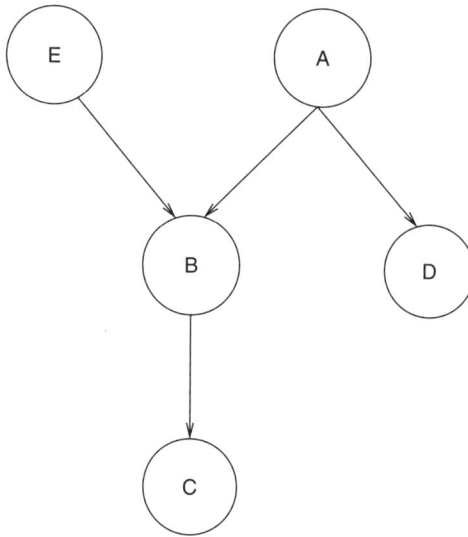

Figure 8.1   Example of a simple network structure.

Hence, in Bayesian networks, the conditional probability of x_i given all its predecessors in the graph is equal to the conditional probability of x_i given only the Markovian parents of x_i, denoted by $Pa(x_i)$. We can express the joint probability of p network variables as a product of conditional probabilities as

$$P(x_1, \ldots, x_p) = \prod_i P\{x_i|Pa(x_i)\}. \tag{8.1}$$

This fact facilitates the economical representation of joint distributions in Bayesian networks, since the entire joint distribution table does not need to be stored. This property is the core of many search algorithms for learning Bayesian networks from data.

On the other hand, the global Markov property summarizes all conditional independencies embedded by the directed acyclic graph by identifying the Markov blanket of each node. The Markov blanket of a node is given by the parents, the children, and the parents of the children of that node which are required to specify the complete conditional distribution of that node. For example, the members of the Markov blanket of x_E are x_B (child) and x_A (parent of the child). Similarly, the Markov blanket of x_B contains x_E, x_A (parents) and x_C (child). The important result is that x is conditionally independent of other variables (not included in the Markov blanket) conditioned on the variables within the Markov blanket. So x_E is independent of all other variables conditioned on x_B and x_A. This property is the foundation of many algorithms such as Gibbs sampling. Gibbs sampling uses the global Markov property to express the complete conditional distribution of each node x_i given the current values of the other nodes. The general expression for the complete conditional for x_i is

$$P(x_i | \text{all other } x_j, j \neq i) \propto P(x_i | Pa(x_i)) \prod_l P(ch(x_i)_l | Pa(c(x_i)_l)), \quad (8.2)$$

where $Pa(x_i)$ and $ch(x_i)$ are values of the parents and children of x_i and $ch(x_i)_l$ are values of the parents of the lth child of x_i.

8.4 Inference for Network Models

To draw inferences about a Bayesian network based on variables $x_1, \ldots, x_p$, we start with n measurements of these p variables in a data set $D = [\{x_1(1), \ldots, x_p(1)\}, \ldots \{x_1(n), \ldots, x_p(n)\}]$. The data for our applications are gene expression microarrays that simultaneously measure the level of mRNA transcription for the genes. The network model contains two different components for inference based on the data: the graphical structure of conditional dependencies, which is more of a model identification or model selection problem; and estimation of the parameters corresponding to the conditional distributions to quantify the dependence structure. Thus we need to identify the structure of the graphical model and estimate the parameters corresponding to this model. Conditioned on a network model, the posterior inference of the parameters can be done using standard approaches. The more challenging task is to identify the network model which will best fit the data. This problem can be expressed as a model selection problem where we have a model space with several network models $M = \{N_1, \ldots, N_g\}$, where N_i is the ith network model describing a particular dependency structure of $x_1, \ldots, x_p$. Criterion-based methods using the BIC and minimum description length (mdl) are popular due to fast computation algorithms (Spirtes et al., 1993; Lauritzen, 1996; Whittaker, 1990), but methods based on the marginal likelihood or the Bayes factor are more coherent model selection methods in a Bayesian framework (Sebastiani et al., 2003, 2005).

If $P(N_h)$ is the prior probability of the hth network, then using Bayes' theorem we can find the posterior probability of the hth network as

$$P(N_h|D) \propto P(N_h)P(D|N_h), \tag{8.3}$$

which is the evidence of support for the hth network from the data. The difficult part is the computation of the marginal likelihood $P(D|N_h)$ as it contains high-dimensional integration as $P(D|N_h) = \int_{\theta_h} P(D|\theta_h, N_h)P(\theta_h|N_h)d\theta_h$ where θ_h is the vector of parameters corresponding to the jth network model which is possibly of very high dimension. The local Markov property and conjugate structures can be used to reduce the complexity of this computation. The local Markov properties encoded by the network N_h imply that the joint distribution of the nodes for the lth sample can be written as

$$P(D_l|\theta_h) = \prod_{i=1}^{p} P(x_i(l)|Pa(x_i(l)), \theta_h), \tag{8.4}$$

where $D_l = \{x_1(l), \ldots, x_p(l)\}$ is the data from the lth sample, $Pa(x_i(l))$ is the configuration of the parents of x_i, and θ_h is the set of parameters for the network N_h. By assuming exchangeability of the data, that is, samples are conditionally independent given the model parameters, the likelihood is given by

$$P(D|\theta_h) = \prod_{l=1}^{n}\prod_{i=1}^{p} P(x_i(l)|Pa(x_i(l)), \theta_h). \tag{8.5}$$

Choice of convenient local priors for θ_h can make the procedure computationally efficient, though it may depend on the structure of the likelihood. One approach is the use of priors following the directed hyper-Markov law (Dawid and Lauritzen, 1995). Under this assumption, the prior density $P(\theta_h)$ admits the same factorization of the likelihood function, hence can be written as $P(\theta_h) = \prod_{i=1}^{p} P(\theta_{hi})$, where θ_{hi} is the subset of parameters used to describe the dependency of x_i on its parents. Thus the marginal likelihood can be expressed as

$$P(D|N_h) = \prod_{l=1}^{n}\prod_{i=1}^{p}\int_{\theta_{hi}} P(x_i(l)|Pa(x_i(l)), \theta_{hi})P(\theta_{hi})d\theta_{hi} = \prod_{i} P(D|N_{hi}), \tag{8.6}$$

where $P(D|N_{hi})) = \prod_l \int_{\theta_{hi}} P(x_i(l)|Pa(x_i(l)), \theta_{hi})P(\theta_{hi})d\theta_{hi}$, the likelihood contribution of the ith node. To complete the hierarchical prior specifications, we need to specify priors for the networks. We may obtain a further simplification by assuming decomposable network prior probabilities which can be expressed as $P(N_h) = \prod_{i=1}^{p} P(N_{hi})$ (Heckerman et al., 1995). Using this prior, the posterior can be expressed in a product form as $P(N_h|D) = \prod_{i=1}^{p} P(N_{hi}|D)$ where $P(N_{hi}|D)$ is the posterior probability weighting the dependency of x_i on the set of of parents specified by the model N_{hi}. The basic assumption here is that the prior probability of a local structure N_{hi} is independent of the other local dependencies N_{hk}, where $i \neq k$. One choice of this local prior probability is $P(N_{hi}) = (a + 1)^{-1/b}$ where $a + 1$ is the cardinality of the model space and b is the cardinality of the set of variables.

Simpler assumptions of these types have a huge impact in the computational efficiency and induce local computation. To compare networks that differ for the parent structure of x_i, we only need to compute the local marginal likelihood corresponding to x_i. Therefore, two local network structures N_h and N_t that specify different parents for the variable x_i can be compared by evaluation of local posterior odds as $P(N_{hi}/D)/P(N_{ti}|D) = P(D|N_{hi})/P(D|N_{ti}).P(N_h)/P(N_t)$. Furthermore, with the assumption of uniform priors, maximization of the marginal likelihood at individual nodes can provide information about the local model.

Parametric inference given a network can be done using the posterior distribution of the parameter:

$$P(\theta_h|D) = \frac{P(D|\theta_h)P(\theta_h)}{P(D|N_h)} = \prod_{i=1}^{p} \frac{P(D|\theta_{hi})P(\theta_{hi})}{P(D|N_{hi})}.$$

It is clear that conjugate choices of prior $P(\theta_h)$ with respect to the likelihood $P(D|\theta_h)$ allow us to obtain the local marginal likelihoods explicitly and thus produce computationally efficient algorithms. The next two examples will be based on these conjugate models.

8.4.1   Multinomial-Dirichlet Model

In this situation we use gene expression values discretized into two categories, 0 and 1, depending on whether the genes are similar to (not expressed) or different (expressed) than the respective control. One possible finer discretization involves three categories, -1, 0, and 1, depending on whether the expression level is significantly lower than (negatively expressed), similar to (not expressed), or greater than the respective control. The control expression level of a gene can be either determined experimentally (as in the method of DeRisi et al., 1997) or set as the average expression level of the gene across the experiment. The meaning of 'significantly' is determined by setting a threshold on the ratio between measured expression and control.

In general, suppose the variables $x_1, \ldots, x_p$ are all discrete and can take values from a_i categories. The multinomial distribution will be a natural local model for the conditional distributions. Suppose that, for the ith node, the conditional probability that x_i is in the kth category conditioned on the event that the parent of x_i, $Pa(x_i)$, is in the jth setup is $P(x_i = k|Pa(x_i) = j) = \theta_{ijk}$.

Figure 8.2 describes a simple network where x_3 depends on x_1 and x_2. Each x_i is binary (so $a_i = 2$) taking values 1 or 2 (true/false, expressed/not

x_1	
1	2
θ_{11}	θ_{12}

x_2	
1	2
θ_{21}	θ_{22}

case k	x_1	x_2	x_3
\multicolumn{4}{}{Database of 10 cases}			
1	1	1	2
2	1	2	1
3	2	1	1
4	1	2	2
5	1	1	1
6	1	2	1
7	2	2	1
8	1	2	2
9	1	1	2
10	2	2	1

			x_3	
$Pa(x_3)$	x_1	x_2	True	False
$Pa(x_3)_1$	1	1	θ_{311}	θ_{312}
$Pa(x_3)_2$	1	2	θ_{321}	θ_{322}
$Pa(x_3)_3$	2	1	θ_{331}	θ_{332}
$Pa(x_3)_4$	2	2	θ_{341}	θ_{342}

			x_3	
$Pa(x_3)$	x_1	x_2	1	2
$Pa(x_3)_1$	1	1	1	2
$Pa(x_3)_2$	1	2	2	2
$Pa(x_3)_3$	2	1	1	0
$Pa(x_3)_4$	2	2	2	0

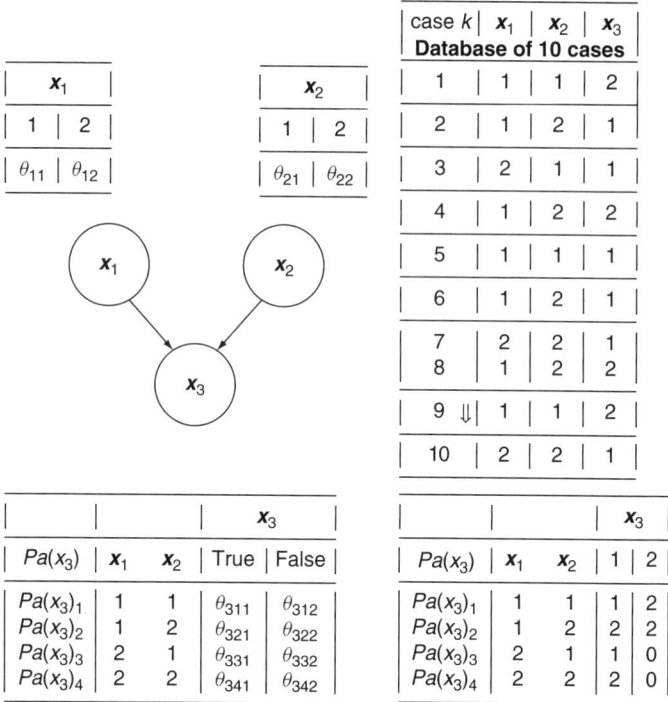

Figure 8.2 A simple Bayesian network describing the dependency of x_3 on x_1 and x_2, which are marginally independent. The table on the left describes the parameters θ_{3jk} ($j = 1, \ldots, 4$ and $k = 1, 2$) used to define the conditional distributions of $x_3 = x_{3k} | Pa(x_3)_j$, assuming all variables are binary. The two tables on the right describe a simple database of seven cases, and frequencies n_{3jk}. The full joint distribution is defined by the parameters θ_{3jk}, and the parameters θ_{1k} and θ_{2k} that specify the marginal distributions of x_1 and x_2.

expressed). The table clarifies the parameters corresponding to the model. For example, θ_{11} represents the probability that x_1 is in the first category while θ_{12} is the probability that x_1 is in the second category. θ_{312} denotes that x_3 is in the second category given that x_1 and x_2 (the parents) are in the first configuration, which is $x_1 = 1$ and $x_2 = 1$.

Now the data set contains n_{ijk} which is the sample frequency of the joint occurrence of $x_i = k$ and $Pa(x_i = j)$. For example, n_{312} denotes the frequency that x_3 is in category 2 given x_1 and x_2 are in the first configuration (i.e. $x_1 = 1$ and $x_2 = 1$), and the value is $n_{312} = 2$. We can calculate the marginal frequencies as $n_{ij} = \sum_k n_{ijk}$, which is the marginal frequency of $Pa(x_i = j)$. From the table we can calculate n_{11}, which is the number of cases where x_1 is in the 1st category and $n_{11} = 7$. Similarly, $n_{21} = 4$. We can calculate all other frequencies in similar fashion by counting the cases. Using these frequencies we can write the multinomial likelihood function as

$$P(D|N) \propto \{\theta_{11}^7 \times \theta_{12}^3\}\{\theta_{21}^4 \times \theta_{22}^6\}\{\theta_{311}^1\theta_{312}^2 \times \theta_{321}^2\theta_{322}^2 \times \theta_{331}^1\theta_{332}^0 \times \theta_{341}^2\theta_{342}^0\}.$$

In the general setup the likelihood function will be

$$P(D|\theta_h) = \prod_{ijk} \theta_{ijk}^{n_{ijk}}.$$

The hyper-Dirichlet distribution with parameters α_{ijk} is the conjugate hyper-Markov law and the density function is $P(\theta_{ijk}) \propto \prod_{ijk} \theta_{ijk}^{\alpha_{ijk}-1}$. The hyperparameters α_{ijk} usually satisfy the consistency rule $\sum_j \alpha_{ij} = \alpha$ (Geiger and Heckerman, 1997) for all i, so that the parameters θ satisfy global and local parameter independence (Spiegelhalter and Lauritzen, 1990). One of the simple choices is $\alpha_{ijk} = \alpha/(c_i l_i)$ where l_i is the number of states of the parent x_i which is known to obey the symmetric Dirichlet distribution. With this choice, as we have a multinomial-Dirichlet conjugate form, we can easily derive the marginal likelihoods for the network N as

$$\prod_{i=1}^p P(D|N_{hi}) = \prod_{ij} \frac{\Gamma(\alpha_{ij})}{\Gamma(\alpha_{ij} + n_{ij})} \prod_k \frac{\Gamma(\alpha_{ijk} + n_{ijk})}{\Gamma(\alpha_{ijk})},$$

where Γ denotes the gamma function.

8.4.2   Gaussian Model

Discretization of the expression measurements may lead to loss of information. Another approach is to exploit the continuous expression data to build the network model. Now the variables $x_1, \ldots, x_p$ are all continuous and it is assumed that the conditional distribution of each variable x_i given its parent $Pa(x_i) = \{x_{i1}, \ldots, x_{ip(i)}\}$ follows a Gaussian distribution with mean μ_i and variance σ_i^2. The local dependence structure has been created through the conditional mean structure as a linear function of the parent variables

$$\mu_i = \beta_{i0} + \sum_{j=1}^{p_i} \beta_{ij} x_{ij}. \tag{8.7}$$

In this way, the dependency x_i on the parent x_{ij} is equivalent to having the regression coefficient $\beta_{ij} \neq 0$.

Suppose we have n i.i.d. samples of observations and $x_i = \{x_{i1}, \ldots, x_{in}\}$. The design matrix Z_i is of dimension $n \times p(i) + 1$, where the lth row (corresponding to the lth sample) is given by $(1, x_{i1l}, x_{i2l}, \ldots, x_{ip(i)l})$ and β denotes the $(Pa(i) + 1) \times 1$ vector of regression coefficients $(\beta_{i0}, \beta_{i1}, \ldots, \beta_{i Pa(i)})$. The linear model can be expressed as

$$x_i = Z_i \beta + \epsilon_i,$$

where ϵ is random error with mean 0 and variance $\sigma^2 I_n$. Based on this model, the likelihood can be written as

$$P(D|\theta) \propto \prod_{i=1}^{p} \tau_i^{n/2} \exp\{-\tau_i(x_i - Z_i\beta_i)'(x_i - Z_i\beta_i)/2\}.$$

The next step is to assign conjugate priors for β_i and τ_i to obtain the marginal likelihood explicitly.

The conjugate prior for β_i, τ_i is NIG, where we assume $\beta_i|\tau_i \sim N(\beta_{i0}, (\tau_i R)^{-1})$ and $\tau \sim \text{Gamma}(\alpha_{i1}, \alpha_{i2})$. We assume R to be an identity matrix.

Under this specification the marginal likelihood is available explicitly as

$$P(D|N) \propto \prod_i \frac{\det R_{i0}^{1/2} \Gamma(\nu_{i*}/2)(\nu_{i0}\sigma_{i0}^2)^{\nu_{i0}/2}}{\det R_{i*}^{1/2} \Gamma(\nu_{i0}/2)(\nu_{i*}\sigma_{i*}^2)^{\nu_{i*}/2}}, \tag{8.8}$$

where the updated parameters after observing the data are given by

$$R_{i*} = R_{i0} + Z_i'Z_i,$$

$$\beta_{i*} = R^{-1}{}_{i*}(R_{i0}\beta_{i0} + Z_i'x_i),$$

$$\alpha_{i1*} = \nu_{i0}/2 + n/2,$$

$$\alpha_{i2*} = \{(-\beta_{i*}'R_{i*}\beta_{i*} + x_i'x_i + \beta_{i0}'R_{i0}\beta_{i0}) + 1/\alpha_{i2}\}^{-1},$$

$$\nu_{i*} = \nu_{i0} + n,$$

$$\sigma_{in} = 2/(\nu_{i*}\alpha_{i2*}).$$

8.4.3   Model Search

The conjugate prior and local modeling allow us to calculate the marginal likelihood values explicitly. However, the space of possible sets of parents for each variable grows exponentially with the number of candidate parents, and efficient search procedures are needed to develop a network model successfully. One simple way is to define a set of candidate parents for each variable x_i, then implement an independent model selection for each variable x_i, and link together the local models selected for each variable x_i. The K2 algorithm (Cooper and Herskovitz, 1992) can further reduce the computation complexity by means of a bottom-up strategy. This strategy starts with a null model (say, one in which x_i has no parents) and evaluates its marginal likelihood. In the next step, a single parent is allowed for each node and the marginal likelihood of each model is evaluated. If the maximal marginal likelihood of these models is larger than the marginal likelihood of the independence model, the parent that increases the likelihood most is accepted and the algorithm proceeds to evaluate models with two parents. If none of the models has marginal likelihood that exceeds that of the independence model, the search stops. The K2 algorithm is implemented in Bayesware Discoverer and the R package deal (Boettcher and Dethlefsen, 2003). The algorithm is based on a greedy search strategy and therefore can be trapped in local maxima and induce spurious dependency. However, it often performs as well as other more complex algorithms (Cooper and Herskovitz, 1992).

Bayesware Discoverer is a knowledge discovery system based on the enabling technology of Bayesian networks. It deploys a unified framework which regards the knowledge discovery process as the automated generation of Bayesian networks from data. The core of Bayesware Discoverer implements a novel methodology to discover Bayesian networks from possibly incomplete databases, a generalization of the well-known Bayesian methodology for learning Bayesian networks from data. Bayesware Discoverer implements several search strategies, and in particular the K2 algorithm. The K2 algorithm works by selecting the, a posteriori, most probable Bayesian network from a subset of all the possible Bayesian networks. The subset of models is selected by the user, who is asked to identify the order in which the variables in the data set are evaluated. The rank of each variable defines the set of variables that will be tested as possible parents: the higher the order of a variable, the larger the number of variables that will be tested as its possible parents. If the user does not specify an order, Bayesware Discoverer uses the order of appearance of the variables in the database to build an initial network that can be further explored to select other dependencies to be tested. The implementation of the K2 algorithm in Bayesware Discoverer starts from the highest ranked variable, say $x1$, and computes, first, the marginal likelihood of the model that assumes no links pointing to $x1$ from the other variables in the list. The next step is the computation of the marginal likelihood of the models with one link only pointing to $x1$ from the other variables in the list. If none of these dependencies has a marginal likelihood larger than that of the model without links pointing to $x1$, the latter is taken as most probable model and the next variable in the list is evaluated. If at least one of these models has a marginal likelihood larger than that of the model without links pointing to $x1$, the corresponding link is accepted and the search continues by trying adding two links pointing to $x1$, and so on, until the marginal likelihood does not increase any more. Once the evaluation of one variable is terminated, the algorithm removes the variable $x1$ from the list by replacing it with the second variable in the original list and repeats the same search. The fact that data are complete, as no entries are reported as unknown in the data set, is a key feature in keeping the induction of Bayesian networks computationally feasible. When data are incomplete under an ignorable missing data mechanism, the marginal likelihood of a Bayesian network becomes a mixture of marginal likelihood induced from the possible completions of the data, with the consequent loss of decomposability properties. Bayesware Discoverer implements a method based on bound and collapse to compute a first-order approximation of the posterior probability of a Bayesian network for the missing data case. As the current version of Bayesware Discoverer handles discrete variables only, continuous variables are discretized into a number of bins that can be chosen by the user and there are two discretization methods available. Continuous variables can be discretized either into a number of equal-length bins, or into a number of bins having approximately the same frequency of cases.

The alternative search procedure is to impose some restrictions on the search space and use the deomposibility of the posterior probability of the network. One way is to create some ordering on the variables such as $x_i > x_j$ if x_i cannot be a parent of x_j. In this way the search space is reduced to a subset of networks in which only a subset of directed associations exist. The ordering information can be induced using known pathway information (Segal et al., 2003). The ordering operation can be largely automated by using programs such as MAPPFinder or GenMAPP. Other more extensive search strategies are based on genetic algorithms (Larranaga et al., 1996), stochastic methods (Singh and Valtorta, 1995), or MCMC methods (Friedman and Koller, 2003).

It is very common in practice to interpret Bayesian networks to represent causal relationships. Actually parent–child relationships in a derived network need not be causal. Furthermore, an edge between variables in the network does not imply a direct biological mechanism. There may exist several intermediate variables between the linked variables that may not have been included in the model. Any network developed from the gene expression data should be biologically validated. Derivation of a biological explanation for the links may not be possible, but, based on the learning about the biological interactions between these genes and their products, we can interpret the significance of the edges that occur in this network. New insights into the types of biological inference can thus be gained from the network.

8.4.4   Example

The gene expression data used in this study result from a study of 31 melanoma samples. For that study, total mRNA was isolated directly from melanoma biopsies, and fluorescent cDNA from the message was prepared and hybridized to a microarray containing probes for 8150 cDNAs (representing 6971 unique genes). Several analytical methods were applied to the expression probes from well-measured genes to visualize the overall expression pattern relationships among the 31 cutaneous melanoma tumor samples. A statistical measure was employed to generate a gene list weighted according to the gene's impact on minimizing the volume occupied by the groups and maximizing center-to-center inter-group distance. The ten most important genes were selected for the]break analysis:

G1 PHO-C
G2 WNT5A
G3 PIRIN
G4 MART-1
G5 S100
G6 HAHDB
G7 STC2
G8 RET-1
G9 SYNUCLEIN
G10 MMP3

Figure 8.3   Loading the data.

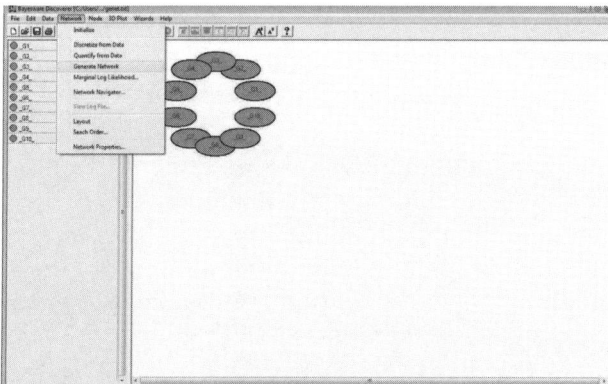

Figure 8.4   Screen with unconnected network.

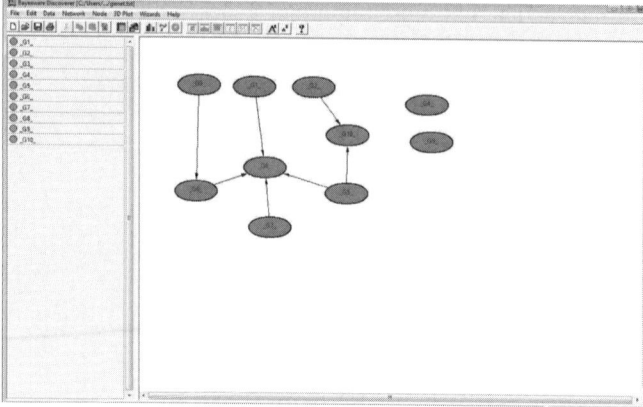

Figure 8.5   Generated Network.

These ten genes were selected for analysis on the basis of either their known or likely roles in the WNT5A-driven induction of an invasive phenotype in melanoma cells, or their close predictive relationships between the genes.

To use Bayes Discoverer to analyze these data, the first step is to load data with a header column containing node names and place data in the columns, as in Figure 8.3. After loading the data, a screen appears with all the nodes unconnected, as in Figures 8.4. Then select **Network** and click on **Generate Network** option which produces the network based on these data as shown in Figure 8.5.

9

Advanced Topics

9.1 Introduction

In previous chapters, we concentrated on modeling aspects of microarray data which conform to some standard form of data collection, in order to answer relevant biological and scientific questions such as differential expression of genes, classification using gene expression profiles, and identification of new subtypes of diseases using clustering approaches. While most scientific questions are addressed by the toolkit we have already described in detail, sometimes we observe data from microarray experiments that require new classes of methods to analyze them. Two important examples are time course experiments and studies where the response is time-to-event. We briefly discuss the Bayesian analytic tools available to model such data.

9.2 Analysis of Time Course Gene Expression Data

Monitoring gene expression patterns over time provides key information about the multidimensional dynamics of complex biological systems. In this time series paradigm we are able to observe the emergence of coherent temporal responses of many interacting components. For example, one of the most important ways in which cells regulate gene expression is by using a feedback loop. Some of the proteins regulate the expression of other genes by either initiating or repressing transcription, and they are known as transcription factors (TFs). Cells exposed to a new condition (e.g. infection, stress, starvation) react by activating a new expression program. This program starts by activating a few TFs, which in turn activate many other genes that act in response to the new condition. Snapshot

Bayesian Analysis of Gene Expression Data   B. Mallick, D. Gold, and V. Baladandayuthapani
© 2009 John Wiley & Sons, Ltd

data can only provide information about the expressed genes at that time point. However, to identify the complete set of expressed genes as well as to determine the interactions between these genes, it is necessary to measure gene expression data over time. These time series expression experiments can answer several questions about biological systems such as circadian rhythms (Panda et al., 2002; Storch et al., 2002), genetic interaction and knockouts (Gasch et al., 2000; Zhu et al., 2000), or progression of infectious disease (Nau et al., 2002; Whitfield et al., 2002; Xu et al., 2002).

Time course microarray data differ from static microarray data in that gene expression levels at different time points can be correlated. Additionally, there are very few replicates and time points, but a relatively large number of genes. Hence, standard time series methods may be neither appropriate nor feasible, and to analyze such dependent data we need to develop more complex statistical methods (Bar-Joseph, 2004; Androulakis et al., 2007). One big challenge is to extract the continuous representation (over time) of all genes throughout the course of the experiment. Furthermore, many experiments contain a large number of missing time points in a series for any given gene, making gene-specific interpolation infeasible. Naive imputation for imbalanced data may lead to misleading results. Also, the timing of the biological process may be variable so the rate at which similar underlying processes (such as the cell cycle) unfold can be expected to differ across organisms, genetic variants, and environmental conditions. For these reasons, it is hard to combine time series expression data. The static analysis of microarrays, proposed by Tusher et al. (2001), has recently included an ad hoc approach to deal with time course microarray data where slopes or signal areas from each time course curve are compared.

9.2.1 Gene Selection

Gene selection with time course microarray data is a relatively new area of research. Bar-Joseph et al. (2003) considered a likelihood ratio statistic for comparison of the two B-spline curves represented for the genes. Guo et al. (2003) applied estimating equation techniques to construct a variant of the robust Wald statistic. Park et al. (2003) proposed a two-way ANOVA model with a nonparametric permutation test to identify genes that have different gene expression profiles among experimental groups. Storey et al. (2005) presented a significance method to identify differentially expressed genes under both independent and longitudinal sampling schemes. Tai and Speed (2006) derived a one-sample multivariate empirical Bayes statistic to rank genes in the order of differential expression from replicated microarray time course experiments. Luan and Li (2004) proposed a B-spline model-based method for identifying periodic expressed genes. Yuan et al. (2006) suggested a hidden Markov approach to classify genes based on their temporal expression patterns.

Recently, Chi et al. (2007) proposed a Bayesian hierarchical model for time course gene expression data. They denoted by $\mathbf{y}_{gij}$ the vector of log-transformed gene expression data for the gth gene, ith biological condition, and jth subject.

The vector $\mathbf{y}_{gij}$ contains data across time, and its dimension m_{gij} is determined by the number of observed time points and replication over time, and is allowed to differ across genes, biological conditions, and subjects to accommodate possible missing expression data over time. The model is expressed as

$$\mathbf{y}_{gij} = \mathbf{x}_{gij}\boldsymbol{\beta}_{gi} + \gamma_{ij}I_{m_{gij}} + \boldsymbol{\epsilon}_{gij},\qquad(9.1)$$

where $\mathbf{x}_{gij}$ denotes the $m_{gij} \times p$ design matrix for the jth subject, gth gene, and in the ith biological condition, which may include all the covariates, $I_{m_{gij}}$ is an $m_{gij} \times 1$ vector of ones, $\boldsymbol{\beta}_{gi}$ is a $p \times 1$ vector of regression coefficients specific to the gth gene and ith biological condition, and $\boldsymbol{\epsilon}_{gij}$ is an $m_{gij|} \times 1$ vector of measurement errors. The subsequent prior distributional assumption for $\boldsymbol{\beta}_{gi}$ will induce correlation structure between the expression data (over subjects, time, and replicates) for the gth gene in the ith biological condition. γ_{ij} is a random effect which accounts for the subject-specific deviation from the mean expression of the gene in the ith biological condition.

The prior for $\boldsymbol{\beta}$ induces correlation within and between genes over temporal replicates. The prior distribution is specified as

$$\boldsymbol{\beta}_{gi}|\boldsymbol{\mu}_i, \Sigma_i \sim N_p(\boldsymbol{\mu}_i, \Sigma_i),$$

which is conditionally independent (conditioned on $\boldsymbol{\mu}_i$ and Σ_i). Further, it is assumed that

$$\boldsymbol{\mu}_i \sim N_p(m_i, \Phi_i),$$

so that marginally (integrating out $\boldsymbol{\mu}_i$) the $\boldsymbol{\beta}_{gi}$ will be correlated. It can be shown that, integrating out $\boldsymbol{\mu}_i$, $\text{cov}(\boldsymbol{\beta}_{gi}, \boldsymbol{\beta}_{hi}) = \Phi_i$, for $g \neq h$. The hyperprior Φ_i thus induces a priori correlation between $\boldsymbol{\beta}_{gi}$ and $\boldsymbol{\beta}_{hi}$ for the ith biological condition, which in turn induces correlation between expression levels over different genes. The prior hierarchy for $\boldsymbol{\beta}_{gi}$ not only accounts for correlated gene expression over time but also allows for dependence in expression over different genes, both within and between subjects in the same biological condition. Chi et al. (2007) proposed an empirical Bayesian method of eliciting Φ_i. Conjugate priors are assumed for all other parameters. Based on this model, they proposed gene selection criteria to identify genes that show changes in temporal expression patterns among biological conditions. This is a flexible method which can accommodate data generated from complex designs.

9.2.2   Functional Clustering

Clustering is another useful technique where the aim is to gain insight into those genes that behave similarly over the course of the experiment. By comparing genes of unknown function with profiles that are similar to genes of known function, clues as to function may be obtained. Hence, co-expression of genes is of interest. Most clustering algorithms are based on independent data. While clustering is important, most clustering algorithms treat their input as a vector of independent samples, and do not take into account the temporal relationship between the time points. Thus, these algorithms cannot benefit from the known dependencies among consecutive points.

Recently there have been developments in clustering algorithms for time series expression data. Schliep et al. (2003) present a hidden Markov model based clustering algorithm for time series expression data. Holter et al. (2001) used a dynamic version of the singular value decomposition (SVD) to model the temporal dependence. Qian et al. (2001) used local alignment algorithms to study time-shifted and inverted gene expression profiles and to infer causality from time series profiles. Bar-Joseph et al. (2003) developed an optimally ordered hierarchical clustering tree to identify clusters.

Bayesian clustering methods are attractive due to their ability to identify the number of clusters automatically as well as to account for all sources of uncertainty. Most Bayesian clustering methods start from the linear model structure

$$y_{gt} = f(\beta_g, t) + \epsilon_{gt},$$

where y_{gt} is the log-transformed gene expression value for gth gene at the tth time point (experiment), f is some potentially nonlinear function of β and ϵ is some error process usually modeled as independent and Gaussian. The function f can be written as a linear model using basis function expansion,

$$y_{gt} = x_t \beta_g + \epsilon_{gt} \tag{9.2}$$

where x_t is the design matrix, which is usually a function of t. This is a general framework and most of the existing Bayesian methods follow it by choosing x_t appropriately.

Ramoni et al. (2002) used an autoregressive model, their choice for x_t being the past responses. For example, to use the AR(p) model, the rows of X will be $\{1, y_{g(t-1)}, y_{g(t-2)}, \ldots, y_{g(t-p)}\}$. They developed a probabilistic scoring metric based on marginal likelihood and the use of conjugate priors to obtain explicit expression for this likelihood.

Wakefield et al. (2003) used periodic basis functions in x_t and exploited model-based clustering methods to cluster these periodic functions. Heard et al. (2005) used spline basis functions in x_t and proposed a novel Bayesian version of the hierarchical clustering algorithm based on marginal likelihood.

Ray and Mallick (2006) used wavelets as basis functions, integrating Dirichlet process priors with these wavelets to obtain automatic clustering. Wavelet representations are sparse and can be helpful in limiting the number of regressors. Dimension reduction is inherent in their approach. Furthermore, wavelets have the ability to capture functions of different degrees of smoothness. As an example, the gene expression profiles of yeast cell cycles may occasionally depart from the usual cyclic behavior and these shifts will be overlooked, in general, by the periodic basis model. Similarly, due to induction of some specific hormones or chemicals, gene expression data can show sudden changes, such as temporal singularities. Both spline and periodic basis models will fail to identify them and thus misclassify the genes in the clustering structure. Wavelet-based methods will successfully cluster the time course data in these complex situations.

9.2.3   Dynamic Bayesian Networks

Dynamic Bayesian network (DBN) models are currently the focus of a growing number of researchers concerned with discovering novel gene interactions and regulatory relationships from expression data. DBNs are an extension of Bayesian networks (Chapter 8). The main advantage of DBNs for gene expression data is that, unlike Bayesian networks, which are acyclic, DBNs allow for cycles, which are common in many biological systems. Additionally, DBNs can improve the ability to learn causal relationships by relying on the temporal nature of the data. Murphy and Mian (1999) first proposed the use of DBNs to model time-course microarray data, and Ong et al. (2002) were among the first to implement DBNs for this purpose. Their method used prior biological knowledge for grouping together genes which are co-regulated. This allowed them to reduce the dimension of the problem. Perrin et al. (2003) developed a DBN model containing hidden variables to overcome both biological and measurement noise. Their method is a direct extension of the linear regression model. Kim et al. (2003) demonstrated the usefulness of DBNs. Husmeirer (2003) performed a simulation-based analysis to determine the accuracy of DBNs in analyzing gene expression data. They concluded that while the global network recovered by DBNs is not useful, local structures can be recovered to a certain extent. Beal et al. (2005) used variational Bayesian methods to approximate the posterior quantities in DBNs. It is clear that DBNs are a promising direction for modeling temporal gene expression systems.

9.3   Survival Prediction Using Gene Expression Data

In the past few years, several microarray studies with time-to-event outcomes have been collected. The main goal is to predict the time to event (also called survival time) using gene expressions as explanatory variables. Survival times of patients in clinical studies may shed light on the genetic variation. Sometimes it is possible to relate the expression levels of a subset of genes with the patient survival time. Apart from this, it is interesting to estimate the patient survival probabilities as a function of covariates such as levels of clinical risk. Because the event is not observed for the samples/array, standard regression techniques cannot be applied since they do not (properly) take censoring into account. In particular, in the case of microarrays, analysis is further complicated by the high dimensionality.

A wide variety of methods have been proposed to handle time-to-event microarray data. Perhaps the simplest of these are the univariate Cox proportional hazards (PH) model for each gene and selecting those genes that pass a threshold of significance (Rosenwald et al., 2002). The problem with this approach is that it does not account for correlation between the genes and consequently many highly correlated genes are selected. Another approach first clusters the genes and then fits a Cox model using the average expression level of each cluster as covariate (Hastie et al., 2001). However, this method can be sensitive to the choice of clustering algorithm. Other dimension reduction approaches are based on partial least squares (Nguyen and Rocke, 2002) or principal component analysis (Li and Gui, 2004). These methods are based on linear combinations of genes rather than the original variables to achieve dimension reduction. Most of these methods are done as via a two-step approach where the gene selection is done first and then the selected genes are used for survival prediction, thus ignoring the uncertainty in the gene selection step. In this section we summarize some Bayesian approaches that circumvent this problem.

9.3.1   Gene Selection for Time-to-Event Outcomes

Lee and Mallick (2004) perform gene selection by extending the work of George and McCulloch (1993) to a non-Gaussian mixture prior framework. A 'small n, large p' situation is considered for selection and survival analysis with Weibull regression or the PH model. They use of a random residual component in the model consistent with the belief that there may be unexplained sources of variation in the data, perhaps due to explanatory variables that were not recorded in the original study. Sha et al. (2006) consider the accelerated failure time (AFT) model as an alternative to the Cox model for modeling the survival data.

9.3.2   Weibull Regression Model

First we consider the parametric model where the independently distributed survival times for n individuals, $t_1, t_2, \ldots, t_n$ follow a Weibull distribution with parameters α and γ_i^*. The data for the jth individual in the experiment, in addition to his survival time t_j, consists of an observation vector $(y_j, X_{j1}, \ldots, X_{jp})$ where y_j indicates the binary or multicategory phenotype covariate and the p $X_{j1}, \ldots, X_{jp}$ are gene expression vectors. A reparameterized model is more convenient ($\lambda = \log \gamma^*$), whereby

$$P(t|\alpha, \lambda) = \alpha t^{\alpha-1} \exp(\lambda - \exp(\lambda)t^\alpha), \quad t, \alpha > 0.$$

The hazard and survival function follow as $h(t|\alpha, \lambda) = \alpha t^{\alpha-1} \exp \lambda$ and $S(t|\alpha, \lambda) = \exp(-t^\alpha \exp \lambda)$. For the n observations at times $t = (t_1, \ldots, t_n)$, let the censoring variables be $\delta = (\delta_1, \ldots, \delta_n)'$, where $\delta_i = 1$ indicates right censoring and $\delta_i = 0$. The joint likelihood function for all the n observations is

$$L(\alpha, \lambda|\theta) = \prod_{i=1}^{n} P(y_i|\alpha, \lambda_i)^{\delta_i} S(y_i|\alpha, \lambda_i)^{(1-\delta_i)} \tag{9.3}$$

$$= \alpha^d \exp\left\{ \sum_{i=1}^{n} \delta_i \lambda_i + \delta_i(\alpha_i) \log y_i - y_i^\alpha \exp \lambda_i \right\},$$

where $\lambda = (\lambda_1, \ldots, \lambda_n)$ and $\theta = (n, t, \delta)$. In a hierarchical step, a conjugate prior can be induced as $\lambda_i = X_i'\beta + \epsilon_i$, where X_i are gene expression covariates, $\beta = (\beta_1, \ldots, \beta_n)$ is a vector of regression parameters for the p genes, and $\epsilon_i \sim N(0, \sigma^2)$.

A Gaussian mixture prior for β enables us to perform a variable selection procedure. Define $\gamma = (\gamma_1, \ldots, \gamma_p)$, such that $\gamma_i = 1$ or 0 indicates whether the gene is selected or not ($\beta_i = 1$ or 0). Given γ, let $\beta_\gamma = \{\beta_i \in \beta : \beta_i \neq 0\}$ and X_γ be the columns of X corresponding to β_γ. Then the β_γ have distribution $N(0, c(X_\gamma' X_\gamma)^{-1})$, where c is a positive scalar. The indicators γ_i are assumed to be a priori independent with $P(\gamma_i = 1) = \pi_i$, which are chosen to be small to restrict the number of genes.

The hierarchical model for variable selection can be summarized as

$$(T_i | \alpha, \lambda_i) \sim \text{Weibull}(\alpha, \lambda_i). \quad \alpha \sim \text{Gamma}(\alpha_0, \kappa_0),$$

$$(\lambda_i | \beta, \sigma) \sim N(X_i'\beta, \sigma^2), \ (\beta_i | \gamma_i) \sim N(0, \gamma_i c \sigma^2), \quad \sigma^2 \sim IG(a_0, b_0),$$

$$\gamma_i \sim \text{Bernoulli}(\pi_i).$$

As the posterior distributions of the parameters are not of explicit form, we need to use MCMC-based approaches, specifically Gibbs sampling alongside Metropolis algorithms to generate posterior samples. The unknowns are $(\lambda, \alpha, \beta, \gamma, \sigma^2)$ and separate Metropolis–Hastings steps are used to sample λ_i, the indicators γ_i (one at a time), $\varphi = \log \alpha$ and finally (β, σ^2) is drawn as $(\beta_\gamma, \sigma^2 | \gamma, \lambda) \sim NIG(V_\gamma X_\gamma' \lambda, V_\gamma; a_1, b_1)$ where a_1, b_1 are defined as in O'Hagan and Forster (2004). For the details of the conditional distributions, the reader is referred to Lee and Mallick (2004).

9.3.3 Proportional Hazards Model

The Cox PH model (Cox, 1972) represents the hazard function $h(t)$ as

$$h(t|x) = h_0(t) \exp W,$$

where $h_0(t)$ is the baseline hazard function and $W = x'\beta$, where β is a vector of regression coefficients. The Weibull model in the previous section is a special case of the PH model with $h_0(t) = \alpha t^{\alpha-1}$. Kalbfleisch (1978) suggests the nonparametric Bayesian method for the PH model. Lee and Mallick (2004) apply a Bayesian variable selection approach to this model. As before, they assume that $W_i = x_i'\beta + \epsilon_i$, where $\epsilon_i \sim N(0, \sigma^2)$, in which x_i forms the ith column of the design matrix X.

Let T_i be an independent random variable with conditional survival function

$$P(T_i \geq t_i | W_i, \Lambda) = \exp\{-\Lambda(t_i) \exp W_i\}, \quad i = 1, \ldots, n.$$

Kalbfleisch (1978) suggested a gamma process prior for $h_0(t)$. The assumption is $\Lambda \sim \mathcal{GP}(a\Lambda^*, a)$, where Λ^* is the mean process and a is a weight parameter about the mean. If $a \approx 0$, the likelihood is approximately proportional to the partial likelihood, and if $a \uparrow \infty$, the limit of the likelihood is the same as the likelihood when the gamma process is replaced by Λ^*. The joint unconditional marginal survival function is directly derivable as

$$P(T_1 \geq t_1, \ldots, T_n \geq t_n | W) = \exp\{-\sum_i \Lambda(t_i)e^{W_i}\},$$

where $W = (W_1, \ldots, W_n)$. The likelihood with right censoring is then

$$L(W|\theta) = \exp\left\{-\sum_i a B_i \Lambda^*(t_i)\right\} \prod_{i=1}^n \{a\lambda^*(t_i) B_i\}^{\delta_i},$$

where $A_i = \sum_{l \in R(t_i)} \exp W_l$, $i = 1, \ldots, n$, $R(t_i)$ is the set of individuals at risk at time $t_i - 0$, $B_i = -\log\{1 - \exp W_i/(a + A_i)\}$, and $\theta = n, t, \delta)$ as before. The hierarchical structure is as follows:

$$W|\beta_\gamma \sim N(X_\gamma \beta_\gamma, \sigma^2 I), \qquad \beta_\gamma \sim N(0, c\sigma^2 (X_\gamma' X_\gamma)^{-1}),$$

$$\sigma^2 \sim IG(a_0, b_0), \qquad \gamma_i \sim \text{Bernoulli}(\pi_i)$$

The MCMC simulation that follows is similar to the Weibull regression model.

9.3.4 Accelerated Failure Time Model

Sha et al. (2006) consider the accelerated failure time model as an alternative to the Cox model for modeling survival data. Rather than assuming a multiplicative effect on the hazard functions as in the Cox regression, AFT models assume a multiplicative effect on the survival times. The general form of an AFT model is

$$\log(T_i) = \alpha + X_i \beta + \epsilon_i, \quad i = 1, \ldots, n,$$

where $\log(T_i)$ is the log-survival time, α is the intercept term, X_i is the p-dimensional vector of gene expression profiles, and the ϵ_i are i.i.d. errors. Kalbfleisch and Prentice (1980) give a comprehensive treatment of parametric AFT models, and Wei (1992) reviews inference procedures for nonparametric models.

Sha et al. (2006) work with parametric AFT models under normal and t-distribution assumptions for ϵ_i. The censoring is handled via the data augmentation procedure of Tanner and Wong (1987) to impute censored values. Specifically, let $\mathbf{w} = (w_1, \ldots, w_n)^T$, where $w_i = \log(t_i)$ is the augmented data with the censored data imputed. Under Gaussian assumptions for the errors, the T_i follow a lognormal distribution or, equivalently, w_i follows a normal distribution with mean $\alpha + X_i \beta$. Gene selection is then achieved via a mixture prior on β as discussed in Chapter 3 and Section 9.3.3. Note that this representation affords convenient marginalization of the regression parameters which results in a much faster and more efficient MCMC sampler. See Sha et al. (2006) for technical details of the method.

Appendix A

Basics of Bayesian Modeling

A.1 Basics

This appendix introduces Bayesian methods and basic parametric models. As this is an applied text, we avoid going into the theoretical properties of subjective probability in great depth. There is a large literature discussing rational decision making through a set of reasonable axioms which would naturally motivate us to use the Bayesian paradigm For more details, see de Finetti (1937, 1963, 1964), DeGroot (1970), Berger (1985), Savage (1972) and Bernardo and Smith (1994). First we introduce de Finetti's representation theorem (de Finneti, 1930) which provides a link between subjective probability and parametric models. It also introduces the predictive interpretation of parameters and provides a way to make predictive inference.

Let us recall the gene expression example. We wish to determine how the gene expression measurements on a patient were related to the presence or absence of cancer. Say, $\mathbf{x}_i$ represents the vector of gene expression measurements for the ith patient and y_i is the response defined as 1 if the patient has cancer and 0 otherwise, for n patients.

For simplicity, let us start with the situation with no measurement on the predictor for any individual (absence of $\mathbf{x}_i$) and consider the problem of making a prediction for a new response y_{n+1}. Thus the data set is given by $D = \{y_1, \ldots, y_n\}$ and we wish to determine $P(y_{n+1}|D)$, the density of the new response y_{n+1} conditioned on the observed responses D. For gene expression, this involves predicting the probability that a new patient has cancer before making gene expression measurements and given only the binary responses of

Bayesian Analysis of Gene Expression Data B. Mallick, D. Gold, and V. Baladandayuthapani
© 2009 John Wiley & Sons, Ltd

other patients in the study. We use conditional probability to evaluate it as

$$P(y_{n+1}|y_1, \ldots, y_n) = \frac{P(y_1, \ldots, y_n, y_{n+1})}{P(y_1, \ldots, y_n)}. \tag{A.1}$$

To evaluate this conditional probability we require the joint density functions in the numerator and the denominator of equation (A.1). The general representation theorem of de Finetti (1930) advocated us for these choices of joint density function.

A.1.1   The General Representation Theorem

The concept of exchangeability is a key idea in the representation theorem which assumes that the subscripts or labels for identifying the individual observable are noninformative. A sequence of random variables $y_1, \ldots, y_n$ is said to be exchangeable if the joint density,

$$P(y_1, y_2, \ldots, y_n) = P(y_{\pi(1)}, y_{\pi(2)}, \ldots, y_{\pi(n)}),$$

for all permutations π defined on the subscripts $\{1, \ldots, n\}$. In our example this means that whichever individual we label at the first, the second, etc., does not alter the joint density of all the responses $P(y_1, \ldots, y_n)$. This concept of exchangeability can be straightforwardly extended to infinite sequences of random variables and is used in the following theorem.

General Representation Theorem: If $y_1, y_2, \ldots$ is an infinitely exchangeable sequence of real-valued random quantities, then there exista a probability distribution function F over Ξ, the space of all distribution functions, such that the joint distribution $P(y_1, \ldots, y_n)$ for $y_1, \ldots, y_n$ has the form

$$P(y_1, \ldots, y_n) = \int_\Xi \prod_{i=1}^n P(y_i|F) dQ(F),$$

where $P(y_i|F)$ is the density of y_i given that it is distributed according to F, and $Q(F) = \lim_{n \to \infty} P(F_n)$, where $P(F_n)$ is a distribution function evaluated at the empirical distribution function defined by $F_n(y) = \frac{1}{n} \sum_{i=1}^n I(y_i \le y)$. This theorem tells us that we can decompose the joint density of the y_i by conditioning on a distribution F and then integrating over the range of F, Ξ the space of all distribution functions. Note that the distribution Q (known as the prior distribution) encodes our beliefs about the empirical distribution F_n.

When F is completely unspecified, we need to integrate over the space of all distribution functions to obtain the joint density of the responses. This is known as nonparametric modeling (Walker et al., 1999; Dey et al., 1998). In this situation we wish to draw some inference about the unknown F (see Chapter 7). In contrast to nonparametric modeling, we assume that F follows some parametric distribution with finite vector of parameters $\theta \in \Theta$. As F is completely specified by θ, the general representation theorem can be used to express the joint density of the responses as

$$P(y_1, \ldots, y_n) = \int_\theta \prod_{i=1}^n \{P(y_i|\theta)\} P(\theta) d\theta. \tag{A.2}$$

It is clear that to completely specify a Bayesian model we need to specify $P(y_i|\theta)$ (the data distribution which is the basis of the likelihood function) and $P(\theta)$ (the prior distribution of the parameters). Throughout the book we mainly concentrate on parametric models following (A.2), though in Chapter 7 we discuss nonparametric models precisely in the context of clustering distributions.

A.1.2 Bayes' Theorem

In the parametric modeling framework using equation (A.2) we obtain the predictive density of a new response as

$$P(y_{n+1}|D) = \int_\Theta P(y_{n+1}|\theta) P(\theta|D) d\theta, \tag{A.3}$$

where $D = (y_1, \ldots, y_n)$ and

$$P(\theta|D) = \frac{P(D|\theta)P(\theta)}{P(D)} \tag{A.4}$$

$$= \frac{\prod_{i=1}^n P(y_i|\theta)P(\theta)}{\int_\Theta \prod_{i=1}^n P(y_i|\theta)P(\theta)}. \tag{A.5}$$

$P(y_{n+1}|D)$ is known as the posterior predictive distribution, and to evaluate it we need $P(\theta|D)$ which is known as the posterior distribution of θ. Equation (A.4) is obtained by using Bayes' theorem and can be simplified as

$$P(\theta|D) \propto P(D|\theta)P(\theta) \tag{A.6}$$

or

Posterior $\propto$ Likelihood $\times$ Prior.

Hence, the posterior distribution is found by combining the prior distribution $P(\theta)$ with the probability of observing the data given the parameters or the likelihood $P(D|\theta)$.

A.1.3   Models Based on Partial Exchangeability

In most complicated applications it is hard to justify the assumption of exchangeability. A simple example is when a set of cancer patients under study are being treated with two different types of drugs. Responses from the patients under the same drug can be assumed exchangeable, but not patients under both drugs. Therefore, we assume that responses from the patients treated by same drug are exchangeable, and they are independent otherwise. This is known as a partially exchangeable structure. The situation can be generalized for more than one group, assuming within-group observations are exchangeable and between-group observations are independent.

Suppose there are m such groups and $\mathbf{y}_i = y_{i1}, \ldots, y_{in_i}$, $i = 1 \ldots, m$, are the observations from the ith group. Under the assumption of partial exchangeability the observations $y_{i1}, \ldots y_{in_i}$ are exchangeable so the representation theorem can be used and the $\mathbf{y}_i$ are independent of the $\mathbf{y}_j$ for $i \neq j$. Using the assumption of within-group exchangeability and between-group independence, the representation theorem for the joint distribution of the observations will be

$$P(\mathbf{y}_1, \ldots, \mathbf{y}_m) = \int_\theta \prod_{i=1}^m \prod_{j=1}^{n_i} P(y_{ij}|\theta_i)P(\theta), \qquad (A.7)$$

where $\theta_i = (\theta_1, \ldots, \theta_p)$. The proof for the representation theorem for partial exchangeable sequences is given in Bernardo and Smith (1994).

A class of models based on the partial exchangeable structure is the clustering or partition models. Briefly, the method can be thought of as partitioning the sample space, to give d subsets $S_1, \ldots, S_d$. We assume exchangeability only for the data within the same partition S_i, while data associated within different subsets are taken to be independent. Typically there will be several partitions, g out of all possible partitions G, whose relative plausibilities are described by a prior probability mass function $P(g)$. These types of models are described in Chapter 7 in the context of cluster modeling.

A.1.4   Modeling with Predictors

So far we have concentrated on a simple modeling situation without using any predictors or covariates. However, we have additional data such as gene expression levels for each patient, and we need to model the dependence between the responses $y_1, \ldots, y_n$ and their corresponding measurements on the predictor variables $\mathbf{x}_1, \ldots, \mathbf{x}_n$. In our example, $\mathbf{x}$ are the gene expression measurements which can be used for cancer prediction. We express the predictive distribution as

$$P(y_1, \ldots, y_n|\mathbf{x}_1, \ldots, \mathbf{x}_n) = \int_\Theta \left\{ \prod_{i=1}^n P(y_i|\theta(\mathbf{x}_i)) \right\} P(\theta(\mathbf{x}))d\theta(\mathbf{x}). \qquad (A.8)$$

A simple example is the linear regression problem where we assume $P(y_i|\boldsymbol{\theta}(\mathbf{x}_i)) = N(\theta(\mathbf{x}_i), \sigma^2)$ and $\theta(\mathbf{x}_i) = \beta_0 + \beta_1 x_i$. This relationship can be extended in general linear and nonlinear frameworks and is described in Chapter 2.

In our example, Y_i is a binary variable and the binary classification model is suitable for this situation. We use the binary distribution $P(y_i|\boldsymbol{\theta}(\mathbf{x}_i)) = \text{Binary}(\theta(\mathbf{x}_i))$, where $\theta(\mathbf{x}_i)$ is the canonical parameter and $\theta(\mathbf{x}_i) = \beta_0 + \beta_1 x_i$. In this binary model, $\theta(\mathbf{x}_i)$ is related to the mean parameter $\mu(\mathbf{x}_i)$ by

$$\theta(\mathbf{x}_i) = \log[\frac{\mu(\mathbf{x}_i)}{(1 - \mu(\mathbf{x}_i))}].$$

Linear and nonlinear classification models are discussed in Chapter 5.

A.1.5   Prior Distributions

The basic philosophical difference between classical and Bayesian statistical inference concerns the use of prior information or beliefs. Bayesian statisticians contend that the investigator's prior information and beliefs are themselves relevant data and should be considered, along with more objective data, in making inference. This difference regarding the use of prior information is related to a disagreement concerning the definition and interpretation of the concept of probability. Bayesian statisticians redefine the classical concept of probability as long-range relative frequency in terms of probability as a degree of confirmation or belief. For axiomatic formulation of subjective probabilities see DeGroot (1970), de Finetti (1964), and Bernardo and Smith (1994).

Typically prior distributions are specified based on information gathered from past studies or from the opinions of experts. There is a growing literature on methodology for the elicitation of prior information (Craig et al., 1998; Kadane and Wolfson, 1998; O'Hagan, 1998), which brings together ideas from statistics and perceptual psychology.

In order to streamline the elicitation process and lessen the burden of computation, we restrict our prior distribution to familiar families of distributions. Further, to obtain closed-form answers for posterior distributions people assign conjugate priors which are conjugate to the likelihood, that is, which lead to a posterior distribution belonging to the same family as the prior.

The prior distribution $P(\boldsymbol{\theta})$ of $\boldsymbol{\theta}$ should be constructed based on our prior information about $\boldsymbol{\theta}$, but objective priors are also popular among practitioners. The most familiar objective priors are noninformative or default prior distributions. One of the natural choices in this group is the uniform prior. For the univariate parameter θ, for example, when the parameter space is bounded this becomes the uniform distribution over the space. When we have an unbounded parameter space, say $(-\infty, \infty)$, we define the uniform prior as constant over the space as

$$P(\theta) = c, \quad -\infty < \theta < \infty,$$

where c is any positive constant. This prior distribution is improper as $\int_{-\infty}^{\infty} P(\theta)d\theta = \infty$. Still, the posterior distribution could be proper as long as the likelihood has a finite integral with respect to θ as

$$P(\theta|\mathbf{y}) = \frac{P(\mathbf{y}|\theta) \cdot c}{\int_{-\infty}^{\infty} P(\mathbf{y}|\theta) \cdot cd\theta},$$

and the integral of $P(\theta|\mathbf{y})$ is 1 under the integrability condition over the likelihood.

Another limitation of the uniform prior is that it is not invariant under reparameterization. The famous Jeffreys prior (Jeffreys, 1946) handles this problem. For the univariate case with parameter θ the Jeffreys prior is

$$P(\theta) = \{I(\theta\}^{1/2}$$

where $I(\theta)$ is the expected Fisher information in the model, namely

$$I(\theta) = -E_{\mathbf{y}|\theta}\left[\frac{\delta^2}{\delta\theta^2} \log P(\mathbf{y}|\theta)\right].$$

In the multiparameter case Jeffreys prior is defined as

$$P(\boldsymbol{\theta}) = \{|I(\boldsymbol{\theta}^{1/2}|\},$$

where $|\cdot|$ denotes the determinant and $I(\boldsymbol{\theta})$ is the expected Fisher information matrix, with (i, j)th element

$$I(\boldsymbol{\theta}) = -E_{\mathbf{y}|\boldsymbol{\theta}}\left[\frac{\delta^2}{\delta\theta_i\delta\theta_j} \log P(\mathbf{y}|\boldsymbol{\theta})\right].$$

The other convenient choice of priors is conjugate priors (conjugate to the likelihood) where the posterior distribution could be obtained in closed form. In this case the posterior distribution $P(\boldsymbol{\theta}|\mathbf{y})$ belongs to the same family as the priors $P(\boldsymbol{\theta})$. Morris (1983) showed that exponential families, a large family containing most of our known likelihood functions, do actually have conjugate priors, so this is a useful approach in practice.

There are several other methods for constructing priors, both informative and noninformative; see Berger (1985) for an overview.

There is controversy regarding the use of prior distribution $P(\boldsymbol{\theta})$ based on subjective probability in Bayesian analysis. Bayesian inference corresponds essentially to the type of inductive reasoning on which the scientific method is based. The development of any area of science consists of successive revisions of hypotheses, which are often at first stated vaguely. It is characteristic of such a development that in the beginning many diverse hypotheses are proposed but, as evidence accumulates, these hypotheses are revised until relatively few remain and these tend to agree, at least in some major respects. The revision of vague diverse theories as evidence accumulates seems to be the only reasonable way for science to proceed; in a sense, criticism of the Bayesian use of subjective probabilities is no more justified than criticism of initial vague personal theories in science.

A.1.6   Decision Theory and Posterior and Predictive Inferences

Statistical decision theory is a general mathematical framework concerned with optimal actions that one may take when faced with uncertainty. All sorts of statistical inference procedures are encompassed by the theory. Central to the theory is the notion of risk, and this forms the basis of comparison of decision rules. Before discussing this framework we need some basic definitions.

(a) *States of nature*: Unknown states of the world $\omega \in \Omega$. Examples are the parameter θ or a future value of the data $\mathbf{y}_{\mathrm{new}}$.

(b) *Actions*: a in the action space A. Actions correspond to particular decisions that the investigator may take. The action space may be somewhat general. In an estimation problem, the action is to report an estimate of θ, so $A = \Theta$. In confidence interval construction, the action space is the set of intervals within the parameter space, and in hypothesis testing the action space is the set of hypotheses under consideration. In a prediction problem, it is the predictive space.

(c) *Utility function*: $U(\omega, a)$ is a function assigning numerical reward to the investigator if he/she takes action a when ω is the true state of nature. Technically it is a mapping from the product space $\Omega \times A$ into the real line $\mathbf{R}$, and we usually restrict attention to nonnegative utilities. Instead of working with $U(a, \omega)$, people work with a so-called *loss function* $L(a, \omega) = -U(a, \omega)$. Examples of popular loss functions are squared-error loss $(\omega - a)^2$, absolute error loss $|\omega - a|$, and zero–one loss where $L(\omega, a) = 1$ if $\omega = a$ and $L(\omega, a) = 0$ otherwise.

(d) *Decision rules*: $\delta(\mathbf{y})$, which is a mapping from the sample space Y into the action space A. In other words, it is a way of associating an action with any data; i.e., it is a statistic. Depending on the inference problem, $\delta(\mathbf{y})$ may be an estimator, a confidence interval, a test statistic or a predictive value.

(e) *Belief distribution*: $P(\omega)$, a specification, in the form of a probability distribution, of current beliefs about the possible states of the world. Examples are $P(\theta) = $ initial beliefs about a parameter vector θ, $P(\theta|\mathbf{y}) = $ beliefs about θ, given data $\mathbf{y}$, and $P(\mathbf{y}_{\mathrm{new}}|\mathbf{y}) = $ beliefs about future data $\mathbf{y}_{\mathrm{new}}$, given data $\mathbf{y}$.

(f) *Risk*: The risk $R(\omega, \delta)$ is the expected loss incurred by the investigator who uses decision rule δ when ω is the state of nature. Suppose we are concerned with the inference problem where $\omega = \theta$. Now we get different types of risks depending on how we are computing the expectation. If we compute the expectation relative to the sampling model, this is known as *frequentist risk*, $\int_Y L(\theta, \delta(y)) P(y|\theta) dy$. The expectation can be taken with respect to the prior distribution of θ as $\int_\theta L(\theta, \delta(\mathbf{y})) P(\theta) d\theta$, which is known as *prior risk*. In our context, the inference occurs after the data $\mathbf{y}$ have been observed so the required distribution of θ is the posterior distribution $P(\theta|\mathbf{y})$. If we take the expectation with respect to this posterior distribution we get the *posterior risk* $\int_\theta L(\theta, \delta(\mathbf{y})) P(\theta|\mathbf{y}) d\theta$. Similarly, for prediction problems the expectation is taken over the marginal distribution of the data.

(g) *Bayes risk*: The Bayes risk is the marginal expected loss, averaging over both data and parameter space, which is $\int_\theta \int_{\mathbf{y}|\theta} L(\theta, \delta(\mathbf{y})) P(\mathbf{y}|\theta) d\mathbf{y} d\theta$. Now we can choose a rule minimizing the Bayes risk which is known as the *Bayes rule*. The name is confusing as a proper Bayesian would not average over the data to choose the decision rule, but instead minimize the posterior risk. However, it turns out that these two operations are virtually equivalent. Under very general conditions, minimizing the Bayes risk is equivalent to minimizing the posterior risk (Berger, 1985).

(h) *Point estimation*: In the problem of point estimation the decision consists of asserting a single value as an estimate of θ. For parametric point estimation $A = \Theta$, the parameter space, and $L(\theta, a)$ are specified. Under squared-error loss the Bayes rule can be obtained by minimizing the posterior risk,

$$R(\theta, a) = \int (\theta - a)^2 P(\theta|\mathbf{y}) d\theta.$$

To minimize the loss function, we need to take the derivative with respect to a:

$$\frac{\delta}{\delta a}[R(\theta, a)] = \int 2(\theta - a)(-1) P(\theta|\mathbf{y}) d\theta.$$

Setting this expression equal to zero and solving for a, we obtain

$$a = \int \theta P(\theta|\mathbf{y}) d\theta = E(\theta|\mathbf{y}),$$

the posterior mean as solution. To check the other condition of positivity of the second derivative is straightforward. Similarly, it can be shown that under absolute error loss the Bayes estimate is the posterior median, while under zero–one loss the Bayes estimate is the posterior mode.

The posterior mean of $h(\theta)$ (where h is any known function of θ) can be obtained as

$$E\{h(\theta|\mathbf{y})\} = \int h(\theta) P(\theta|\mathbf{y}) d\theta$$
$$= \frac{\int h(\theta) P(\mathbf{y}|\theta) P(\theta)}{\int P(\theta|\mathbf{y}) P(\theta) d\theta}. \tag{A.9}$$

Similarly, we can perform predictive inference constructing proper loss functions and the point estimates will be the predictive mean, median or mode depending on the choice of the loss function.

(i) *Credible region*: Rather than point estimation (a single value of θ) we may be interested about a range of values of θ, the Bayesian analog of the frequentist confidence interval. Therefore, the decision set A is the set of subsets of Θ. The loss function is $L(\theta, a) = I(\theta \notin a) + k \times \text{volume}(a)$, where I is the indicator function. In this situation the Bayes rule will be the region or regions for θ having highest posterior density and k will control the tradeoff between the volume of a and the posterior probability of coverage.

Now to construct a $100(1 - \alpha)\%$ credible set for θ we choose a set A such that $P(\theta \in A|\mathbf{y}) = 1 - \alpha$ (which says that the probability that θ lies in A given the observed data $\mathbf{y}$ is at least α). The set is called a *highest posterior density* credible region if $P(\theta|\mathbf{y}) \geq P(\Phi|\mathbf{y})$ for all $\theta \in A$ and $\Phi \notin A$ which is our Bayes rule under the previously mentioned loss function. Such a credible set is very appealing since it covers the most likely θ values. Similarly, we can construct predictive intervals by replacing the posterior distribution by the corresponding predictive distribution.

(j) *Hypothesis testing*: Now consider a null hypothesis $H_0 : \theta \in \theta_0$ versus an alternative hypothesis $H_1 : \theta \in \theta_0^c = \theta_1$. We can describe this as a decision problem with $\Omega = (\omega_0, \omega_1)$ where $\omega_0 = [H_0 : \theta \in \theta_0]$ and $\omega_1 = [H_1 : \theta \in \theta_1]$ with $\theta \in \Theta = \theta_0 \cup \theta_0^c$. We assign the believe distribution $P(\omega)$ as $p_0 = P(H_0)$ and $p_1 = P(H_1)$, the prior probabilities of the hypotheses. The action space $A = \{a_0, a_1\}$, where a_i is the action corresponding to rejection of the hypothesis H_i for $i = 0, 1$ and the loss function $L(a_i, \omega_j) = L(a_i, H_j) = l_{ij}$, $i, j \in \{0, 1\}$, with l_{ij} reflecting the relative seriousness of the four possible consequences, and typically, $l_{00} = l_{11} = 0$.

Suppose our loss function is $l_{ij} = 1$ if $i = j$ and $l_{ij} = 0$ if $i \neq j$ for $i, j \in \{0, 1\}$. Now the expected loss of the decision a_i (choosing the ith hypothesis H_i for $i = 0, 1$), given $\mathbf{y}$, is thus

$$\int L(a_i, \omega) P(\omega|\mathbf{y}) = \int L(a_i, H) P(H|\mathbf{y}) = P(H_i|\mathbf{y}).$$

The optimal decision is therefore to choose the hypothesis which has the highest posterior probability. Now to compare two hypotheses, a simple measure is the posterior odds, the ratio of the posterior probabilities corresponding to the two hypotheses, which is

$$\frac{P(H_0|\mathbf{y})}{P(H_1|\mathbf{y})} = \frac{P(\mathbf{y}|H_0)}{P(\mathbf{y}|H_1)} \times \frac{P(H_0)}{P(H_1)}$$

where $P(\mathbf{y}|H_i) = \int p_i(\mathbf{y}|\boldsymbol{\theta}_i) p_i(\boldsymbol{\theta}_i) d\boldsymbol{\theta}_i$ is the integrated likelihood (or the marginal likelihood).

Another measure is the Bayes factor,

$$B = \frac{\text{Posterior odds ratio}}{\text{Prior odds ratio}} = \frac{P(H_0|\mathbf{y})/P(H_1|\mathbf{y})}{P(H_0)/P(H_1)}$$

which measures whether the data $\mathbf{y}$ have increased or decreased the odds of H_0 relative to H_1. Thus $B > 1$ signifies that H_0 is more relatively plausible considering $\mathbf{y}$. Good (1988) has suggested using the logarithm of this Bayes factor as the weights of evidence in favor of the hypothesis. We discuss the Bayes factor in detail in Section A.2 in the model choice context.

A.1.7   Predictive Distributions

Often the goal of the statistical analysis is to predict future observations, which is straightforward in this Bayesian setup. Suppose that $\mathbf{y}_{\text{new}}$ are the future observations which are independent of $\mathbf{y}$ given the underlying $\boldsymbol{\theta}$. Then the prior predictive is the marginal distribution (with respect to the prior distribution) $f(\mathbf{y}_{\text{new}}) = \int f(\mathbf{y}_{\text{new}}|\boldsymbol{\theta}) P(\boldsymbol{\theta}) d\boldsymbol{\theta}$, and the posterior predictive is the marginal distribution with respect to the posterior distribution $f(y_{\text{new}}|\boldsymbol{\theta}) = \int f(y_{\text{new}}|\boldsymbol{\theta}) P(\boldsymbol{\theta}|\mathbf{y}) d\boldsymbol{\theta}$.

A.1.8   Examples

Example A.1
Suppose that $Y_1, \ldots, Y_n$ are binary outcomes coded as $Y_i = 1$ for diseased tissues and $Y_i = 0$ for normal tissues as described in Section 5.4.5. Now $Y = \sum_{i=1}^{n} Y_i$ conditional on θ is assumed to have a binomial likelihood

$$P(y|\theta) = \binom{n}{y} \theta^y (1 - \theta)^{n-y}, \quad 0 < \theta < 1.$$

The corresponding conjugate prior for this binomial likelihood is a beta distribution with parameters (α, β):

$$P(\theta) = \frac{\theta^\alpha (1 - \theta)^\beta}{B(\alpha, \beta)}, \quad 0 < \alpha, \beta < \infty.$$

We can also find out the noninformative prior using the binomial likelihood. The posterior density will be proportional to the product of the prior and the likelihood,

$$P(\theta|y) \propto \theta^{\alpha+y-1}(1-\theta)^{\beta+n-y-1}.$$

This is a beta density (so the posterior has the same distribution as that of the prior) with updated parameters $\alpha^* = \alpha + y$ and $\beta^* = \beta + n - y$. Using the standard properties of the beta distribution,

$$P(\theta|y) = \frac{\theta^{\alpha+y-1}(1-\theta)^{\beta+n-y-1}}{Beta(\alpha+y, \beta+n-y)}.$$

Thew prior expectation of θ is

$$E(\theta) = \frac{\alpha}{\alpha+\beta},$$

while the posterior expectation is

$$E(\theta) = \frac{\alpha^*}{\alpha^*+\beta^*} = \frac{\alpha+y}{\alpha+\beta+n}.$$

We can write

$$E(\theta|y) = \frac{ny/n + (\alpha+\beta)(\frac{\alpha}{\alpha+\beta})}{n+(\alpha+\beta)}.$$

Now the sample estimate of θ is the sample proportion $\hat{p} = y/n$, and $\lambda = \alpha + \beta$ is known as the prior sample size. The posterior expectation is

$$E(\theta|Y) = \frac{n\hat{p} + \lambda E(\theta)}{n+\lambda},$$

which is a weighted average of sample proportion $\hat{p}$ and the prior mean of θ. The weights are the actual sample size n and the prior sample size $\lambda = \alpha + \beta$.

Now, after observing y, we wish to predict a new observation y_{new}, which is independent of y, given θ, and possesses a binomial distribution, with probability θ and sample size m. Then the prior predictive probability mass function of y_{new} is

$$P(y_{new}) = \int_0^1 P(y_{new}|\theta)P(\theta)d\theta$$

$$= \frac{\binom{m}{y_{new}}}{B(\alpha,\beta)} \int_0^1 \theta^{\alpha-1}(1-\theta)^{m-y_{new}+\beta-1}d\theta$$

$$= \frac{\binom{m}{y_{new}}B(y_{new}+\alpha, m-y_{new}+\beta)}{B(\alpha+\beta)}, \quad y_{new} = 0, 1, \ldots, m.$$

Similarly, the posterior predictive distribution will be

$$
\begin{aligned}
P(y_{new}|y) &= \int_0^1 P(y_{new}|\theta)P(\theta|y)d\theta \\
&= \frac{\binom{m}{y_{new}}}{B(\alpha^*, \beta^*)} \int_0^1 \theta^{\alpha^*-1}(1-\theta)^{m-y_{new}+\beta^*-1}d\theta \\
&= \frac{\binom{m}{y_{new}}B(y_{new}+\alpha^*, m-y_{new}+\beta^*)}{B(\alpha^*+\beta^*)}, \quad y_{new} = 0, 1, \ldots, m.
\end{aligned}
$$

Finding the posterior predictive mean is simple:

$$
\begin{aligned}
E(y_{new}|y) &= E_{\theta|y}[E(y_{new}|\theta)] \\
&= E_{\theta|y}(m\theta) \\
&= m\frac{\alpha^*}{\alpha^*+\beta^*}.
\end{aligned}
$$

The Jeffreys prior form binomial model is again beta with parameters (0.5, 0.5), which is conjugate, too.

Example A.2

$Y_1, Y_2, \ldots, Y_n$ are an independent sample of size n from the normal distribution with mean μ and variance σ^2. Now in this situation the sample mean $\bar{y}$ is sufficient as $P(\theta|y) = P(\theta|\bar{y})$. We know that $\bar{y} \sim N(\mu, \sigma^2/n)$ so the likelihood function is

$$
P(\bar{y}|\theta) = \frac{\sqrt{n}}{\sigma\sqrt{2\pi}} \exp\left(-\frac{n(\bar{y}-\mu)^2}{2\sigma^2}\right), \quad y \in, \mathbf{R}
$$

where $\theta = (\mu, \sigma)$, $\mu \in \circledR$, $\sigma^2 \in \circledR^+$. Looking at the likelihood as a function of μ alone it is clear that the conjugate is again a normal distribution. We can assume that $P(\mu) = N(\mu|\mu_0, \tau^2)$, where μ_0 and τ^2 are known hyperparameters. Now looking at the likelihood function as a function of σ^2 alone, the conjugate prior will be proportional to $(\sigma^2)^a e^{-b\sigma^2}$, which is similar to the inverse gamma distribution. If $1/\sigma^2$ follows a gamma distribution with parameters (a, b) then σ^2 follows an inverse gamma IG(a, b) distribution, with density function

$$
f(\sigma^2|a, b) = \frac{e^{-1/b\sigma^2}}{\Gamma(a)b^a \sigma^{2a+1}}, \quad a > 0, b > 0.
$$

Sometimes we reparameterize the model in terms of $1/\sigma^2$ which is known as the precision and follows a gamma distribution according to this choice of prior for σ^2. Finally, we assume that μ and σ^2 are a priori independent so that $P(\theta) = P(\mu)P(\sigma^2)$.

The conditional posterior distribution of mean μ turns out to be

$$
P(\mu|\sigma^2, \bar{y}) \sim N\left(\frac{(\sigma^2/n)\mu_0 + \tau^2\bar{y}}{\sigma^2/n + \tau^2}, \frac{(\sigma^2/n)\tau^2}{(\sigma^2/n) + \tau^2}\right)
$$
$$
= N\left(\frac{\sigma^2\mu_0 + n\tau^2\bar{y}}{\sigma^2 + n\tau^2}, \frac{\sigma^2\tau^2}{\sigma^2 + n\tau^2}\right),
$$
(A.10)

so we see that this posterior distribution has mean which is a weighted average of the prior mean and the sample value, with weights that are inversely proportional to the corresponding variances (proportional to the corresponding precisions).

The conditional posterior distribution of σ^2 turns out to be

$$
P(\sigma^2|\mu, \bar{y}) = \text{IG}\left(\frac{n}{2} + a, \left[\frac{n}{2}(\bar{y} - \mu)^2 + \frac{1}{b}\right]^{-1}\right).
$$
(A.11)

The marginal distributions of μ and σ^2 are explicitly available in this conjugate framework.

Furthermore, we can use noninformative priors in this problem; a popular one assumes that μ and $\log\sigma$ are locally uniform or, equivalently,

$$
P(\mu, \sigma) \propto \sigma^{-1}.
$$

Using the prior and normal likelihood, the joint posterior distribution is

$$
P(\mu, \sigma|\mathbf{y}) \propto \sigma^n \exp\left(\frac{-\sum_{i=1}^n (y_i - \mu)^2}{2\sigma^2}\right).
$$

Now

$$
\sum_i (y_i - \mu)^2 = \sum_i (y_i - \bar{y})^2 + n(\bar{y} - \mu)^2
$$
$$
= (n-1)s^2 + n(\bar{y} - \mu)^2
$$

where the sample variance is

$$
s^2 = \sum_i \frac{(y_i - \bar{y})^2}{n - 1}.
$$

Therefore, clearly the conditional distribution $P(\mu|\sigma, \mathbf{y}) \sim N(\bar{y}, \sigma^2/n)$. The marginal distribution of σ is

$$
P(\sigma|\mathbf{y}) \propto \sigma^n \exp\left(-\frac{(n-1)s^2}{2\sigma^2}\right),
$$

so σ is distributed a posteriori as a chi distribution, $\sqrt{n-1}s\chi_n^{-1}$.

The marginal distribution of μ is

$$P(\mu|\mathbf{y}) \propto \left[1 + \frac{n(\mu - \bar{y})^2}{(n-1)s^2} \right]^{-n/2}, \tag{A.12}$$

which will be recognized as

$$\frac{\mu - \bar{y}}{s/\sqrt{n}} \sim t_{n-1}.$$

Example A.3 Mixture prior
Suppose that the likelihood is $P(y|\theta)$ for a univariate parameter θ, and assign two-component mixture prior $P(\theta) = \pi P_1(\theta) + (1 - \pi)P(_2(\theta), \ 0 < \pi < 1$, which is a mixture distribution of p_θ and $p_2(\theta)$. Then

$$
\begin{aligned}
P(\theta|y) &= \frac{P(y|\theta)p_1(\theta)\pi + P(y|\theta)p_2(\theta)(1 - \pi)}{\int [P(y|\theta)p_1(\theta)\pi + P(y|\theta)p_2(\theta)(1 - \pi)]} \\
&= \frac{\dfrac{P(y|\theta)p_1(\theta)}{m_1(y)}m_1(y)\alpha + \dfrac{P(y|\theta)p_2(\theta)}{m_2(y)}m_2(y)(1 - \alpha)}{m_y\alpha + m_2(y)(1 - \alpha)} \\
&= \frac{P(y|\theta)p_1(\theta)}{m_1(y)}w_1 + \frac{P(y|\theta)p_2(\theta)}{m_2(y)}(1 - w_1) \\
&= p_1(\theta|y)w_1 + p_2(\theta|y)(1 - w_1),
\end{aligned} \tag{A.13}
$$

where $m_i(y)$ is the marginal distribution with prior p_i, $i = 1, 2$, and the posterior weight is

$$w_1 = \frac{m_1(y)\pi}{m_1(y)\alpha + m_2(y)(1 - \alpha)},$$

which is updated from the prior weight π. Therefore, the posterior is also a mixture distribution. If we use $p_1(\theta)$ and $p_2(\theta)$ as conjugate priors for the likelihood, we obtain the posterior as a mixture of conjugate posteriors in explicit form.

A.2   Bayesian Model Choice

We have defined a Bayes model and representation theorem for an exchangeable sequence of observations. The predictive model typically has a mixture representation which may not be unique. Now the question is, if there IS more than one model, how to choose the best one. Usually Bayesian inference is done using the posterior distributions, whereas model selection is done using the predictive distribution (Box, 1980).

The model choice could be seen as a decision problem whose solution involves model choice or comparison among the alternatives in $M = \{M_i, i \in I\}$. If we assume that the true model is a member of M and our main intention is to choose the right model (not inference or prediction), then we can think that the unknown state of nature is the true model $M_i \in M$ so $\Omega = M$. Now the action space in this problem is $A = M$ with $a = M_i$, $M_i \in M$. The belief distribution $P(\omega)$ will be assigned over the model space by $P(M_i)$, $M_i \in M$. The next step will be to construct a loss function $L(M_i, \omega)$ (or the utility function $U(M_i, \omega)$) and obtain the true model (best model) M^* by minimizing the loss function.

Model choice could be seen as a special case of hypothesis testing, and we can obtain similar results just by switching the notation from hypotheses H_0 and H_1 to models M_i, $i = 1, 2$. As there is no limit on the number of hypotheses that may be simultaneously considered, extension of these concepts to more than two models are straightforward. For simplicity, suppose that we have two candidate models M_1 and M_2 with respective parameter vectors θ_1 and θ_2. Suppose that the likelihood and the prior corresponding to ith model ($i = 1, 2$) are $p_i(\mathbf{y}|\theta_i, M_i)$ (we express the dependence of the likelihood on the model by conditioning on the model M_i) and $p_i(\theta_i|M_i)$. Then the marginal distribution of $\mathbf{y}$ is found by integrating out the parameters,

$$P(\mathbf{y}|M_i) = \int P(y|\theta_i, M_i) p_i(\theta_i|M_i) d\theta_i, \quad i = 1, 2. \qquad \text{(A.14)}$$

Furthermore, $P(M_2|\mathbf{y}) = 1 - P(M_1|\mathbf{y})$.

Now, similarly to Section A.1.6, if we construct a zero–one loss function we obtain a similar type of results. The loss function $L(M_i, \omega)$ is 0 if $\omega = M_i$, and 1 otherwise. The optimal decision is again to choose the model which has the highest posterior probability. Therefore, we can use the Bayes factor to compare these models:

$$B = \frac{P(M_1|\mathbf{y})/P(M_2|\mathbf{y})}{P(M_1/P(M_2)} = \frac{P(\mathbf{y}|M_1)}{P(\mathbf{y}|M_2)}, \qquad \text{(A.15)}$$

the ratio of the observed marginals for the two models, which provides the relative weight of evidence for model M_1 compared to model M_2.

We expect model choice criteria to follow the principle of Ockham's razor, that is, the chosen model should be no more complicated than necessary. The Bayes factor has an automatic razor built in so that it picks up simpler models. When we compare models of different dimensions we usually use criteria based on penalizing the dimension of the model. Use of a Bayes factor is convenient because it has a natural razor within it to penalize overparameterized models.

One criterion based on a penalty for model dimension is Akaike's (1974) information criterion (AIC). Here we choose the model that minimizes

AIC $= -2($log maximized likelihood$) + 2($number of parameters$)$.

The argument for the AIC is based on asymptotic considerations, and it is known that AIC tends to overestimate the number of parameters needed, even asymptotically. This is mainly due to the small penalty term. Therefore, this criterion was modified by Schwarz (1978) to give the Bayesian information criterion,

$$\text{BIC} = -2(\log \text{maximized likelihood})$$
$$+ \log(\text{sample size})(\text{number of parameters}). \qquad \text{(A.16)}$$

and for large sample sizes it approximates -2 times the log of the Bayes factor.

A serious problem with the Bayes factor is that, if we use vague priors so that $P(\theta)$ is improper, then the marginal distribution $f(\mathbf{Y})$ is improper (though not necessarily the posterior $P(\theta|\mathbf{y})$ which makes it impossible to calibrate the Bayes factor, and it loses its interpretation in this situation).

Another problem with Bayes factor occurs in the case of nested models, when it may exhibit 'Lindley's paradox' (Lindley, 1957). Suppose we are comparing the reduced model with a full model and data points support rejection of the reduced model. Now if the prior is proper but sufficiently diffused, it can place a little mass on alternatives so that the denominator of the Bayes factor is much smaller than the numerator, resulting in support for the reduced model in spite of the data. Therefore, the Bayes factor tends to place too much weight on parsimonious selection.

Example A.4

Suppose that for $\mathbf{y} = (y_1, y_2, \ldots, y_n)$ we have two alternative models M_1 and M_2 with $P(M_i) > 0$, $i = 1, 2$ for a normal model with unknown mean μ and known variance σ^2. Now the predictive models are

$$M_1 : p_1(\mathbf{y}) = \prod_{i=1}^{n} N(y_i|\mu_0, \sigma^2), \qquad \mu_0 \text{ known},$$

$$M_2 : p_2(\mathbf{y}) = \int \prod_{i=1}^{n} N(y_i|\mu, \sigma^2) N(\mu|\mu_1, \lambda^2) d\mu, \qquad \mu_1, \lambda^2 \text{ known},$$

where $N(y_i|\mu, \sigma^2)$ is the normal model with mean μ and variance σ^2. M_1 is the simple normal distribution model with mean μ_0 and variance σ^2. M_2 is the complicated model where the mean μ is unknown. We assign a conjugate prior for this unknown μ which is again a normal distribution with known mean μ_1 and variance λ^2. If the sample mean $\bar{y} = (\sum_{i=1}^{n} y_i)/n$ then the Bayes factor B is

$$B = \frac{N(\bar{y}|\mu_0, \sigma^2/n)}{\int N(\bar{y}|\mu, \sigma^2/n) N(\mu|\mu_1, \lambda^2) d\mu}$$

$$= \left(\frac{1/\lambda^2 + n/\sigma^2}{1/\lambda^2}\right)^{1/2} \frac{\exp\{\frac{1}{2}(\lambda^2 + \sigma^2/n)^{-1}(\bar{y} - \mu_1)^2\}}{\exp\{\frac{1}{2}n(\bar{y} - \mu_1)^2/\sigma^2\}}.$$

Now for fixed $\bar{y}$, the Bayes factor $B \to \infty$ as $\lambda^2 \to \infty$ so that evidence in favour of M_1 (the simple model) is overwhelming as the prior variance becomes very large (so prior precision is vanishingly small, creating a larger and more complex model).

A.3   Hierarchical Modeling

Bayesian hierarchical models allow strength to be borrowed among units across a whole data set in order to improve inference. For example, gene expression experiments used by biologists to study fundamental processes of activation /suppression frequently involve genetically modified animals or specific cell lines, and such experiments are typically carried out only with a small number of biological samples. It is clear that this amount of replication makes standard estimates of gene variability unstable. By assuming exchangeability across the genes, inference is strengthened by borrowing information from comparable units.

When the data are exchangeable the representation theorem is given in equation (A.2). However, we said nothing about the joint prior specification $P(\boldsymbol{\theta}) = P(\theta_1, \ldots, \theta_m)$. Subjective beliefs for specific applications are used to specify the joint prior distribution, but some assumption of additional structures may be useful to model the prior. We may assume that the parameters $\theta_1, \ldots, \theta_m$ are exchangeable themselves and modify the representation theorem as

$$P(\mathbf{y}_1, \ldots, \mathbf{y}_m) = \int_{\boldsymbol{\theta}} \prod_{i=1}^{m} \prod_{j=1}^{n_i} P(y_{ij}|\theta_i) P(\theta_i|\phi) P(\phi). \qquad (A.17)$$

The complete model structure has the *hierarchical* form

$$P(\mathbf{y}_1, \ldots, \mathbf{y}_m|\theta_1, \ldots, \theta_m) = \pi_{i=1}^{m} \pi_{j=1}^{n_i} P(y_{ij}|\theta_i)$$

$$P(\theta_1, \ldots, \theta_m|\phi) = \pi_{i=1}^{m} P(\theta_i|\phi)$$

$$P(\phi).$$

Therefore, the stage-wise assumptions of exchangeability create random samples of observations and parameters from different distributions in different hierarchies (stages). At the first stage we get i.i.d. y_{ij} conditioned on θ_i and at the second stage i.i.d. θ_i conditioned on the hyperparameter ϕ. At the final stage the prior distribution of ϕ specifies the belief about it.

This enterprise of specifying a model over several levels is called hierarchical modeling. The proper number of levels varies with the problem, but since we are continually adding randomness as we move down the hierarchy, subtle changes to levels near the top are not likely to have much impact on the bottom or data level.

Shrinkage arises naturally in Bayesian analysis especially in these hierarchical models, as the effect of the prior distribution in general is to pull the likelihood towards the prior and the Bayes estimate will be a combination of the sample estimate and the prior (pulled estimate) and so have more stable and smaller variances than the other competitors. Whenever we have an assumption of closeness or equality of parameter values a priori it will cause shrinkage automatically.

Consider a simple example with observations y_{ij}, $i = 1, \ldots, N$, $j = 1, \ldots, M$, with i indexing the units and j indexing the subjects within the units. We wish to estimate gene-specific means μ_i. Let us assume that the observations from a unit are exchangeable and the observations between units are independent. Due to the assumption of exchangeability, the observations within a unit will be conditionally independent and we further assume that

$$y_{ij}|\mu_j \sim N(\mu_i, \sigma^2),$$

where σ^2 is known.

The maximum likelihood estimate (MLE) of μ_i is the unit-specific sample mean

$$\bar{y}_{i.} = \frac{\sum_{j=1}^{M} Y_{ij}}{M}.$$

Among the advantages of this estimator are that it is simple to compute and it is an unbiased and consistent estimator of μ_i. However, when n_i is small, the estimates are highly variable. In such a situation, one needs to borrow information from other units if they have a latent characteristic common to them, to obtain a more accurate estimate. One such estimator is the pooled mean,

$$\bar{y} = \frac{\sum_{i=1}^{N} \sum_{j=1}^{M} y_{ij}}{NM},$$

where one pretends that all units are identical. The pooled mean is a biased estimate for any particular μ_i, but is preferred over the unbiased estimate because of its smaller variance. In either case we determine how much pooling is permitted: no pooling or complete pooling. However, we can let the data determine how much pooling is required by using a Bayesian hierarchical model. We can then obtain an estimator of μ_i that has smaller total mean squared error across the N estimates of μ_i. This is an example of hierarchical models where the units and the subjects within the units form the hierarchy.

We further assume that the μ_j are exchangeable and develop a hierarchical model

$$y_{ij}|\mu_i \sim N(\mu_i, \sigma^2)$$

$$\mu_i \sim N(\mu, \tau^2).$$

Due to the conjugate structure of the model, the posterior distribution of μ_i conditional on the data is given by

$$\mu_i | y \sim N(\hat{\mu}_i, N/\sigma^2 + 1/\tau^2),$$

where

$$\hat{\mu}_i = (1 - w)\,\bar{y}_i + w\mu.$$

The term w is called the *shrinkage factor*. Thus the posterior mean (shrinkage estimator) of μ_i is the weighted average of the population pooled mean (μ) and sample unit mean ($\bar{y}_i$). We stated earlier that exchangeability does not mean independence. Below, we provide a brief but important proof concerned with independence in hierarchical models.

We wish to obtain the marginal distribution of y_{ij},

$$f(Y_{ij}|\mu) \propto \int f(Y_{ij}|\mu_i)f(\mu_i|\mu)d\mu_i$$

$$= \int \exp\left[-\frac{(y_{ij} - \mu_j)^2}{\sigma^2}\right] \exp\left[-\frac{(\mu_j - \mu)^2}{\tau^2}\right] d\mu_j.$$

It can be easily seen that the marginal density is

$$y_{ij}|\mu \sim N(\mu, \sigma^2 + \tau^2).$$

To show that conditional independence does not imply marginal independence, we derive the marginal correlation between Y_{ij} and Y_{kj} using the law of iterated expectation as follows. We have

$$E(y_{ij}) = E(E(y_{ij}|\mu_i)) = \mu,$$

and the marginal variance is given by

$$\text{var}(y_{ij}) = \text{var}(E(y_{ij}|\mu_i)) + E(\text{var}(y_{ij}|\mu_i)) = \sigma^2 + \tau^2.$$

The covariance between y_{ij} and Y_{ik} is

$$\text{cov}(y_{ij}, y_{ik}) = E(y_{ij}y_{ik}) - E(y_{ij})E(y_{kj})$$

$$= E(E(y_{ij}y_{ik}|\mu_i)) - \mu^2$$

$$= E(\mu_i^2) - \mu^2$$

$$= \tau^2.$$

Then the marginal correlation can be obtained as

$$\text{corr}(Y_{ij}, Y_{kj}) = \frac{\tau^2}{\sigma^2 + \tau^2}.$$

Thus, the observations within a unit are exchangeable but not independent. This is crucial in analysis of gene expression data where the units could be the state of the disease, for example control and cancer (or cancer of different kinds), and the observations with the units are from different genes or signatures. Assumptions of independence may not be valid here as the dependence among genes is a known biological fact. Assumption of exchangeability and development of a hierarchical model can facilitate dependence through simple conditional independence structures.

The usage of the pooled mean allows us to borrow information across all the units. It can be easily shown that, under certain conditions, the mean squared error of $\hat{\mu}_j$ is lower than that of $\bar{Y}_j$. The shrinkage estimator is biased, but makes better use of the total information available to yield superior performance in terms of total mean squared error. If σ^2 is small relative to the total variation, we do not shrink our estimate of μ_j towards the pooled mean μ But if σ^2 is large relative to τ^2, the resulting estimate will shrink towards μ. We demonstrate this idea using two examples.

Example A.5

Here we consider the example given in Goldstein et al. (1993), where the aim is to differentiate between good and bad schools. A sample of 1978 students (subjects) is chosen from 38 schools (units) and their test scores are measured along with several covariates such as gender. We wish to calculate each school's observed average score, and the overall average for all schools. Since some schools have as few as three students, it might be a good idea to model the students as exchangeable given the school information. This allows us to borrow strength across schools to obtain improved estimates of school averages. By way of illustration, we consider a sample without any covariates. The model is

$$y_{ij}|\mu_i \sim N(\mu_i, \sigma_y^2),$$

$$\mu_i \sim N(\mu, \sigma_\mu^2),$$

where y_{ij} is the test score of the jth student from the ith school. We elicit conjugate priors for the unknown parameters (random variables):

$$\sigma_y^2 \sim \text{IG}(u_y, v_y),$$

$$\mu \sim N(\mu_0, \sigma_\mu^2),$$

$$\sigma_\mu^2 \sim \text{IG}(u_\mu, v_\mu).$$

Here, $\text{IG}(u, v)$ is the notation for inverse gamma distribution with mean $v/(u-1)$ and variance $v^2/(u-1)^2(u-2)$, $u > 2$, and $N(\mu, \sigma^2)$ denotes a normal distribution with mean μ and variance σ^2. Figure A.1 shows the 95% confidence and 95% credible intervals of scores for individual schools using the direct sample estimator and the Bayes shrinkage estimator. Note that in

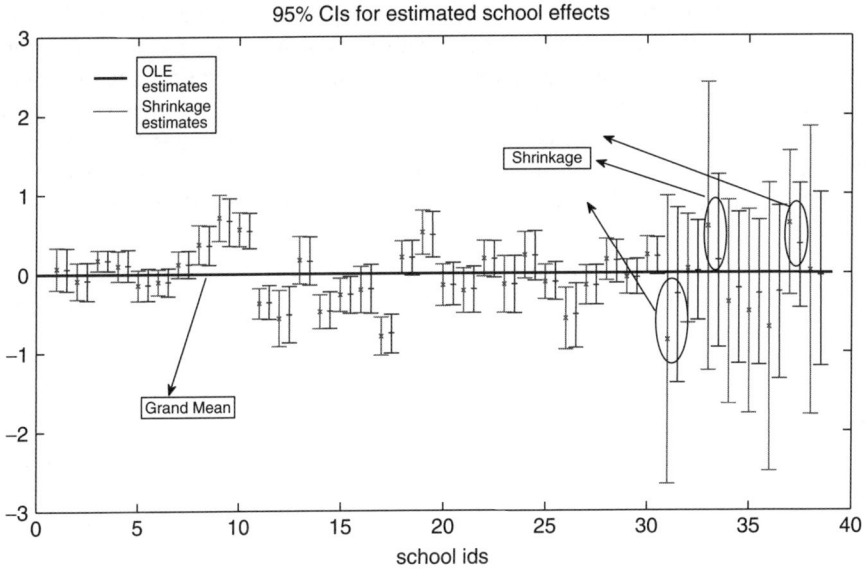

Figure A.1   School effects using maximum likelihood and Bayes estimates.

most cases the shrinkage estimator yields a narrower interval, indicating better precision, as a consequence of borrowing information across schools.

Example A.6
Gelfand et al. (1990) considered modeling rat growth. Weekly measurements of rat weights were taken for five weeks. The response variable, y_{ij}, is the weight of rat i at week j, $i = 1, \ldots, 30$, $j = 1, \ldots, 5$, and the explanatory variable is X_j, the days on which the weights were taken. Each rat has its own expected growth line, but we wish to borrow information across rats to tell us about individual rat growth. The individual expected growth line of rat i is given by

$$E(y_{ij}) = \alpha_i + \beta_i X_j.$$

The average population growth line is given by

$$E[E(Y_{ij})] = \alpha_c + \beta_c X_j.$$

Information across the rats is borrowed using the following hierarchical structure:

$$E(Y_{ij}|\alpha_i, \beta_i) = \alpha_i + \beta_i X_j$$
$$\alpha_i \sim N(\alpha_c, \tau_\alpha)$$
$$\beta_i \sim N(\beta_c, \tau_\beta). \qquad (A.18)$$

The model described above can be interpreted as saying that each rat has its individual growth line but these lines themselves share a common distribution. In other words, the individual growth line models the additional effect of the covariates affecting individual rats over the general effect. The average population growth line essentially models the general effect of the covariates that affect all the rats. The marginal expected growth line is given by $E(Y_{ij}|\alpha_c, \beta_c)$. Using the law of iterated expectation, we get the marginal expected growth line as follows:

$$E(Y_{ij}|\alpha_c, \beta_c) = E(E(Y_{ij}|\alpha_i, \beta_i))$$
$$= E(\alpha_i + \beta_i X_j)$$
$$= \alpha_c + \beta_c X_j.$$

Note that the marginal model follows the average population growth model. This phenomenon can be interpreted as saying that if we integrate out the additional effect of the covariates affecting individual rats, we are left with the general effect of the covariates affecting all the rats. Now a hierarchal model can be written as:

$$y_i \equiv (y_{i1}, y_{i2}, \ldots, y_{in_i})^T$$
$$\theta_i \equiv (\alpha_i, \beta_i)^T$$
$$y_i|\theta_i, \sigma_y^2 \sim N(X_i\theta_i, \sigma^2 I_{n_i}).$$

Here X is the appropriate design matrix and I_J is the identity matrix of size J. The priors for the unknown parameters (random variables) are elicited as

$$\theta_i|\mu_c, \Sigma_c \sim N(\mu_c, \Sigma_c),$$
$$\sigma^2 \sim \text{IG}(u, v,)$$
$$\mu_c \sim N(\eta, C),$$
$$\Sigma_c^{-1} \sim \text{W}((\rho R)^{-1}, \rho),$$

where $W(V, n)$ is the Wishart distribution with mean nV. In Figure A.2(a) the individual growth lines obtained using the MLEs of the regression parameters along with the population average growth line are shown. In Figure A.2(b) Bayes shrinkage estimates of the individual growth lines obtained using the regression parameters along with the population average growth line are shown. Note that the Bayes shrunk growth line shrinks the individual growth lines toward the overall population average.

In summary, using hierarchical models, we can (i) borrow strength from other observations, (ii) shrink estimates toward overall averages, (iii) update the model in the light of available data, and (iv) incorporate prior/other information in estimates and account for other sources of uncertainty.

A.4   Bayesian Mixture Modeling

The general problem of inferring a hypothesis, a model choice, or a complex decision has received much attention in the subjective probability framework. Bayesian mixture modeling offers subtle advantages over many alternatives for decision making, providing a framework capable of incorporating the full problem uncertainty. Bayesian mixture modeling has its roots in frequentist theory, extending the basic mixture modeling, allowing much greater flexibility and control with prior information.

Suppose that observations $\mathbf{y} = (\mathbf{y_1}, \ldots, \mathbf{y_n})$ are distributed according to the likelihood

$$P(\mathbf{y}|\boldsymbol{\theta}, \boldsymbol{\pi}) = \prod_{i=1}^{n} \sum_{k=1}^{K} \pi_k \mathbf{P_k}(\mathbf{y_i}|\theta_k), \tag{A.19}$$

a weighted sum of likelihoods $P_1, \ldots, P_k$, indexed by $\theta_1, \ldots, \theta_K$ with mixture weights $\boldsymbol{\pi} = (\pi_1, \ldots, \pi_K)$. This is an example of a finite mixture model. Infinite mixture models are discussed in Chapter 7. In practice, the parameters $\boldsymbol{\theta}$ and $\boldsymbol{\pi}$ are unobserved, although it is often the case that prior information is available about one or more components. In the case of no prior information, the data informs decisions about the mixture components. Usually when a mixture prior is specified, it results in a posterior that is also a mixture, with weights naturally updated according to the data. Often the mixture is expressed introducing latent variables, z_{ik}, $i = 1, \ldots, n$ and $k = 1, \ldots, K$,

$$P(\mathbf{y}|\boldsymbol{\theta}, \boldsymbol{\pi}) = \prod_{i=1}^{n} \prod_{k=1}^{K} P_k(y_i|\theta_k)^{z_{ik}}, \tag{A.20}$$

where the z_{ik} follow a multinomial distribution with parameters $\boldsymbol{\pi} = (\pi_1, \ldots, \pi_K)$, naturally imposing the constraints $\sum_k z_{ik} = 1$ and $\sum_k \pi_k = 1$. The posterior predictive distribution of a new observation y_{n+1} is also a mixture, derived by integrating over the full posterior uncertainty,

$$
\begin{aligned}
P(y_{n+1}|\mathbf{y}) &= \int_{\Theta} \int_{\Pi} \sum_k \pi_k P_k(y_{n+1}|\theta_k) P(\theta_k, \pi_k|\mathbf{y}) d\boldsymbol{\theta} d\boldsymbol{\pi} \\
&= \sum_{k=1}^{K} \int_{\Theta_k} \int_{\Pi_k} \pi_k P_k(y_{n+1}|\theta_k) P(\theta_k, \pi_k|\mathbf{y}) d\theta_k d\pi_k \\
&= \sum_{k=1}^{K} \alpha_k P_k(y_{n+1}|\mathbf{y}),
\end{aligned}
\tag{A.21}
$$

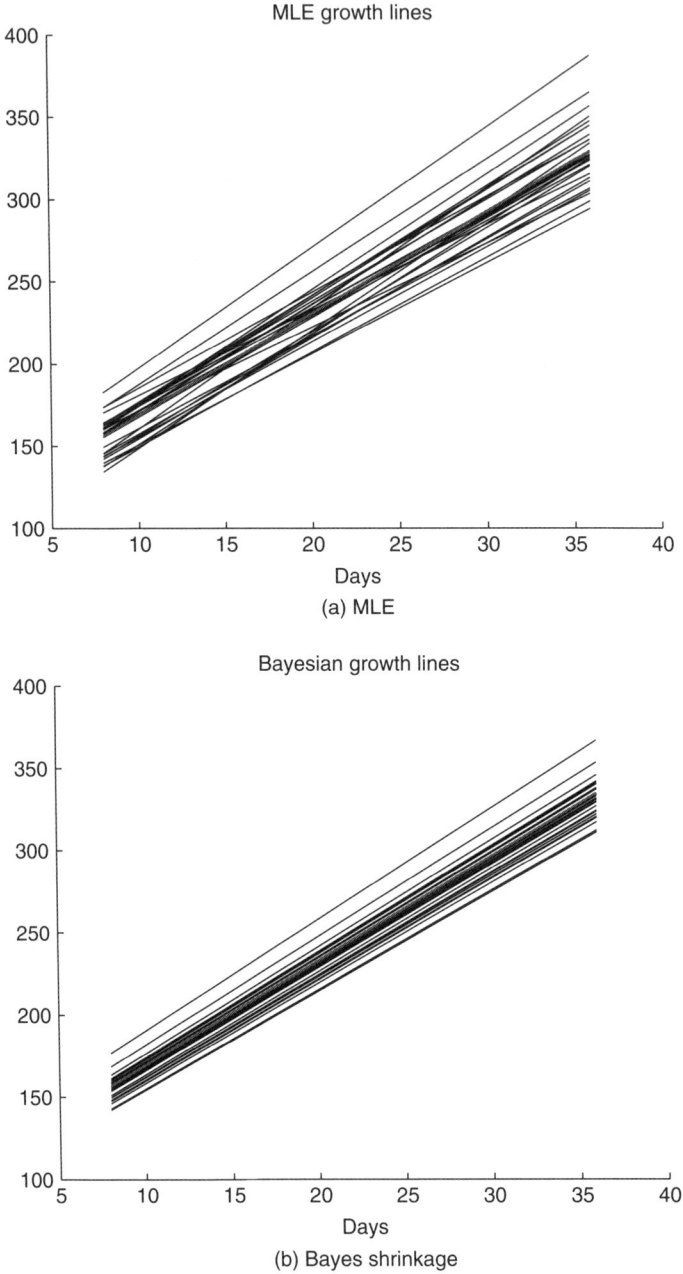

Figure A.2   Growth lines using maximum likelihood and Bayes estimates.

a mixuture of predictive densities, where we let

$$\int_{\Theta_k}\int_{\Pi_k}\pi_k P_k(y_{n+1}|\theta_k)P(\theta_k,\pi_k|\mathbf{y})d\theta_k d\pi_k = \alpha_k\int_{\Theta_k}P_k(y_{n+1}|\mathbf{y},\theta_k)P(\theta_k|\mathbf{y})d\theta_k,$$

$$= \alpha_k P_k(y_{n+1}|\mathbf{y}) \qquad (A.22)$$

for all $k = 1,\ldots,K$. A widely specified prior for $\boldsymbol{\pi}$ is the Dir(α) density, for its conditional conjugacy; in the special case where $K = 2$ this is the beta distribution.

Example A.7
Let y_i, $i = 1,\ldots,n$, be i.i.d. according to the mixture model

$$P(y_i) = \pi 1_0 + (1-\pi)N(0,\sigma^2), \qquad (A.23)$$

an additive mixture of a point mass at 0 and a Gaussian density centered at 0 with unknown scale σ, with probabilities π and $1-\pi$, respectively. The priors of the unobserved parameters π and σ^2 are specified as

$$\pi \sim \text{Beta}(a,b),$$

$$\sigma^2 \sim \text{IG}(v/2, v\sigma_0^2/2). \qquad (A.24)$$

Reexpressing the posterior with the use of latent variables,

$$P(\pi,\sigma^2|\mathbf{y}) \propto \prod_{i=1}^{n}(\pi 1_0)^{z_i}((1-\pi)N(y_i;0,\sigma^2))^{1-z_i}\pi^{z_i}(1-\pi)^{1-z_i}$$

$$\times \pi^{a-1}(1-\pi)^{b-1}\{\sigma^2\}^{-v/2-1}\exp\left\{-\frac{v\sigma_0^2}{2\sigma^2}\right\}, \qquad (A.25)$$

the expression does not admit a recognizable form. In Chapter 7 we discuss Bayesian mixture models in the context of unsupervised classification for gene expression profiles, as well as computational methods for finite mixtures where K is known, and where K is unknown.

A.5   Bayesian Model Averaging

Model choice is a popular approach where we select one model from some class of models and assume that as the true model for the generation of the data. By choosing a single model as correct we ignore model uncertainty and create overconfident inference. As 'none of the models are true but some of them are useful', assumption of a single true model may not be a reasonable approach and a mixture model may capture the unknown better than a single model. In fact this mixing over models also improves the predictive performance.

Suppose that $M_1, M_2, \ldots, M_k$ is our collection of candidate models and $M_i = \{P(\mathbf{y}|\theta_i, M_i), P(\theta_i|M_i)\}$ the likelihood and prior for model i. Let ϕ be the quantity of interest, assumed to be well defined for each model. If the prior model probabilities are $\{P(M_1), \ldots, P(M_k)\}$ then the posterior distribution of ϕ is given by

$$P(\phi|\mathbf{y}) = \sum_{i=1}^{k} P(\phi|M_i, \mathbf{y}) P(M_i|\mathbf{y}), \qquad (A.26)$$

where $P(\phi|M_i, \mathbf{y})$ is the posterior for ϕ under the ith model and $P(M_i|\mathbf{y})$ is the posterior probability of the ith model given by

$$P(M_i|\mathbf{y}) = \frac{P(\mathbf{y}|M_i) P(M_i)}{\sum_{j=1}^{k} P(\mathbf{y}|M_j) P(M_j)}. \qquad (A.27)$$

$P(\mathbf{y}|M_i) = \int P(\mathbf{y}|\theta_i, M_i) P(\theta_i|M_i) d\theta_i$ is the marginal distribution of the data under the ith model. Madigan and Raftery (1994) note that averaging over all possible models in this fashion provides better average predictive ability, as measured by a logarithmic scoring rule, than using a single model. The main difficulty of this method is that the number of potential models k could be extremely large; for example, in a variable selection problem with 15 predictors there will be 2^{15} models. Madigan and Raftery (1994) have used the Ockham's window method to average over parsimonious, data-supported models. An alternative would be to search the entire sample space using MCMC, locating the high-probability models. The problem will be more difficult if we need to search the parameter space simultaneously with the model space, when the marginalization of θ is not analytically possible.

Appendix B

Bayesian Computation Tools

B.1 Overview

Bayesian computation tools are at the heart of Bayesian methods. As the science of statistics has advanced, so have Bayesian methods and the computation tools needed to apply them. These tools, while not unique to science, are generally much more widely employed by Bayesian rather than frequentist statisticians, and arguably have shaped Bayesian thinking as much as Bayesian thinking has shaped science.

While, on the one hand, Bayesian methods offer flexible modeling strategies, there are computational hurdles to consider. Complex dependencies between variables, as well as intricate levels of variation, can be dealt with quite naturally in a Bayesian paradigm, if one is willing to work with and develop the required sophisticated computation algorithms. Many factors need to be considered when choosing a Bayesian computation algorithm, such as the complexity of the posterior distribution, the degree of difficulty of computing posterior probabilities of events of interest, and the probability statements desired, to name but a few.

In this appendix we consider the computation problems of posterior estimation in the spotlight of Bayesian application, and take as given that the likelihood $f(x|\theta)$ and prior $p(\theta)$ distributions are adequate and fully reflect the modeler's beliefs about the experimental design and all sources of variation. The posterior distribution of the parameters of interest is not required to have an analytical closed form, i.e., the normalizing constant can be unknown.

Recall that the normalizing constant is given by $p(x) = \int_{\Theta} p(x|\theta)\pi(\theta)d\theta$. Even in the event that $p(\theta|x)$ does have a closed form, i.e. $p(x)$ is known, the posterior probability distribution might still not admit a familiar form that is convenient to work with, to make for example statements such as $p(\theta \in R|x)$

Bayesian Analysis of Gene Expression Data B. Mallick, D. Gold, and V. Baladandayuthapani
© 2009 John Wiley & Sons, Ltd

for some set R contained in the parameter space, using conventional software. In bioinformatics applications this can very often be the case; we must resort to methods for approximating integrals over the posterior distribution, especially when $p(x)$ is unknown and cannot be evaluated analytically. In high dimensions we can apply strategies based on Monte Carlo sampling or simulation.

B.2   Large-Sample Posterior Approximations

Where they exist, large-sample approximations to posterior distributions, or integrals involving posterior measures, can offer a great deal of convenience. The theoretical elegance of these methods attracts many, owing to a vast body of frequentist work in asymptotics. While posterior approximations can be time-saving devices, their use is limited to applications where an approximation is suitable, to an appropriate order of magnitude. We begin with the fundamental and most straightforward of posterior approximations, provided by the *Bayesian central limit theorem*.

B.2.1   The Bayesian Central Limit Theorem

Like its frequentist counterpart this is a large-sample approximation, requiring large samples for the approximation to work well. Let $x_1, \ldots, x_n$ be independent observations from $f(x|\theta)$, and suppose that $\pi(\theta)$ is the prior for θ. The prior may be improper, as long as the posterior distribution is proper and its mode exists. Then as $n \to \infty$,

$$\theta|x \to N_p \left[\hat{\theta}_m, H^{-1}(\hat{\theta}_m) \right]$$

with posterior mode $\hat{\theta}_m$ and posterior covariance $H^{-1}(\hat{\theta}_m)$. The $p \times p$ matrix

$$H(\theta) = -\frac{\partial^2 \log p^*(\theta|x)}{\partial \theta_i \partial \theta_j}$$

is the negative of the *Hessian matrix* over $p^*(\theta|x) = p(x|\theta)\pi(\theta)$, the kernel of the (unnormalized) posterior density. The estimator of the asymptotic covariance matrix is minus the inverse of the Hessian matrix, evaluated at $\hat{\theta}_m$:

$$H^{-1}(\hat{\theta}_m) = H^{-1}(\theta)|_{\theta=\hat{\theta}_m}.$$

The Bayesian central limit theorem tells us that once the appropriate normal distribution is identified, for a large enough sample size n, the posterior probability of an event E can be approximated by

$$P(E|x) \approx \int_{\theta \in E} N_p \left(\theta; \hat{\theta}_m, H^{-1}(\hat{\theta}_m) \right) d\theta.$$

Of course, the approximation depends on how closely the true posterior distribution resembles a normal distribution, and also on n, being a first-order approximation. The proof of the Bayesian central limit theorem has been discussed extensively, and can be found elsewhere. We direct the interested reader to Carlin and Louis (2000).

Example B.1

Suppose that a cohort of n women is studied to link the development of benign ovarian cysts with gene expression. We assume that the subjects are not exposed to environmental risks of disease that would increase their susceptibility, and are categorized as otherwise healthy individuals. Gene expression measurements $x_i' = (x_{i1}, \ldots, x_{iG})$ for G genes are obtained at time t_0, the beginning of the study. The binary response variables $y_1, \ldots, y_n$ are recorded for n subjects and represent the presence or absence of a benign ovarian cyst during the follow-up period with regular six-monthly visits to the doctor. Responses are assumed independent Binomial($1, \theta_i$) random variables, with success (in this case presence of a cyst) probability θ_i. The logistic function links the θ_i to gene expression,

$$\theta_i = \frac{\exp(x_i'\beta)}{1 + \exp(x_i'\beta)},$$

with unknown β, a $G \times 1$ vector of regression coefficients. The likelihood as a function of β is given by

$$p(y|\beta) = \prod_{i=1}^{n} \theta_i^{y_i}(1 - \theta_i)^{1-y_i}$$

$$= \prod_{i=1}^{n} \left[\frac{\exp(x_i'\beta)}{1 + \exp(x_i'\beta)}\right]^{y_i} \left[\frac{1}{1 + \exp(x_i'\beta)}\right]^{1-y_i}$$

$$= \exp\left\{\sum_{i=1}^{n}\left[y_i x_i'\beta - \log\left(1 + e^{x_i'\beta}\right)\right]\right\}.$$

Setting a a uniform (improper) prior for β, i.e. $\pi(\beta) \propto 1$, it follows that the posterior mode of β is the maximum likelihood estimate, $\hat{\beta}$, since the posterior is proportional to the likelihood

$$p(\beta|y) \propto \exp\left\{\sum_{i=1}^{n}\left[y_i x_i'\beta - \log\left(1 + e^{x_i'\beta}\right)\right]\right\}.$$

The variance–covariance matrix is derived via the Hessian,

$$\frac{\partial}{\partial \beta_j} \log p^*(\beta|y) = \sum_{i=1}^{n} \left(y_i - \frac{\exp(x_i'\beta)}{1 + \exp(x_i'\beta)} \right) x_{ij}$$

$$-\frac{\partial^2}{\partial \beta_j \partial \beta_k} \log p^*(\beta|y) = \sum_{i=1}^{n} x_{ij} x_{ik} \frac{\exp(x_i'\beta)}{\left(1 + \exp(x_i'\beta)\right)^2}$$

$$H(\beta) = -\frac{\partial^2 \log p^*(\beta|x)}{\partial \beta \partial \beta'}$$

$$= X'VX,$$

where V is an $n \times n$ diagonal matrix with ith diagonal element

$$v_{ii} = \frac{\exp(x_i'\beta)}{\left(1 + \exp(x_i'\beta)\right)^2},$$

for $i = 1, \ldots, n$. Asymptotically,

$$\beta|y \rightarrow N_p\left(\hat{\beta}, (X'\hat{V}X)^{-1}\right),$$

where $\hat{V} = V|_{\beta=\hat{\beta}}$ evaluated at $\hat{\beta}$ is

$$\begin{bmatrix} \frac{\exp(x_1'\hat{\beta})}{\left(1+\exp(x_1'\hat{\beta})\right)^2} & & 0 \\ & \ddots & \\ 0 & & \frac{\exp(x_n'\hat{\beta})}{\left(1+\exp(x_n'\hat{\beta})\right)^2} \end{bmatrix}_{n \times n}.$$

B.2.2   Laplace's Method

Another class of posterior approximations are provided by Laplace's method of approximating integrals, which dates from 1774 (Carlin and Lewis, 2000). Laplace's method has been studied extensively, and recent extensions, discussed below, provide second-order convergence, without strong dependence on the Gaussian assumption.

We are interested in approximating integrals (one-dimensional for the time being) of the form

$$\int f(\theta) e^{-nq(\theta)} d\theta,$$

for smooth functions f and q of θ, assuming that q attains a unique minimum at $\hat{\theta}$. It is straightforward to see, expanding the functions f and q as Taylor's series around $\hat{\theta}$, that

$$\int f(\theta)e^{-nq(\theta)}d\theta \approx \int \left[f(\hat{\theta}) + \frac{f'(\hat{\theta})}{1!}(\theta - \hat{\theta}) + \frac{f''(\hat{\theta})}{2!}(\theta - \hat{\theta})^2 \right]$$

$$\times \exp \left[-nq(\hat{\theta}) - nq'(\hat{\theta})(\theta - \hat{\theta}) - nq''(\hat{\theta})\frac{(\theta - \hat{\theta})^2}{2!} \right] d\theta$$

$$= \int e^{-nq(\hat{\theta})} \left[f(\hat{\theta}) + f'(\hat{\theta})(\theta - \hat{\theta}) + \frac{f''(\hat{\theta})}{2}(\theta - \hat{\theta})^2 \right]$$

$$\times \exp \left[-\frac{(\theta - \hat{\theta})^2}{2(nq''(\hat{\theta})^{-1}} \right] d\theta.$$

Notice that the quadratic term in the exponent resembles a normal kernel, forcing the second term in brackets to vanish under integration. The expectation can be approximated by

$$\int f(\theta)e^{-nq(\theta)}d\theta \approx e^{-nh(\hat{\theta})} f(\hat{\theta}) \sqrt{\frac{2\pi}{nh''(\hat{\theta})}} \left[1 + \frac{f''(\hat{\theta})}{2f(\hat{\theta})} \text{ var}(\theta) \right],$$

for $\text{var}(\theta) = \left[nh''(\hat{\theta}) \right]^{-1} = O(n^{-1})$, i.e. a first-order approximation. In p dimensions

$$\int f(\theta)e^{-nq(\theta)}d\theta \approx f(\hat{\theta}) \left(\frac{2\pi}{n} \right)^{p/2} |\tilde{\Sigma}|^{1/2} \exp \left[-nh(\hat{\theta}) \right],$$

for the $p \times p$ matrix

$$\tilde{\Sigma} = \left[\frac{\partial q(\theta)}{\partial \theta_j \partial \theta_k} \right]^{-1} |_{\theta = \hat{\theta}}.$$

A very useful second-order approximation to Laplace's method can be achieved by means of the following ingenious substitution (Tierny and Kadane, 1986). Suppose that we wish to compute the posterior expectation of a function $g(\theta) > 0$. We can let $-nq(\theta)$ equal the unnormalized posterior density,

$$-nq(\theta) = \log[p(x|\theta)\pi(\theta)].$$

The expectation of $g(\theta)$ is

$$E[g(\theta)] = \frac{\int g(\theta)e^{-nq(\theta)}d\theta}{\int e^{-nq(\theta)}d\theta} = \frac{\int e^{-nq^*(\theta)}d\theta}{\int e^{-nq(\theta)}d\theta},$$

with $q^*(\theta) = -\frac{1}{n}\log(g(\theta)) + q(\theta)$. We can now apply Laplace's method as before with $f = 1$ in both the numerator and denominator. The integral is

$$E[g(\theta)] = \frac{|\tilde{\Sigma}^*|^{1/2}\exp\left[-nq^*(\hat{\theta}^*)\right]}{|\tilde{\Sigma}|^{1/2}\exp\left[-nq(\hat{\theta})\right]}\left[1 + O(n^{-2})\right],$$

where θ^* maximizes $-q^*$ and

$$\Sigma^* = \left[\frac{\partial q^*(\theta)}{\partial\theta_j\partial\theta_k}\right]^{-1}|_{\theta=\theta^*},$$

approximated by

$$E[g(\theta)] \approx \frac{|\Sigma^*|^{1/2}\exp\left[-nq^*(\hat{\theta}^*)\right]}{|\Sigma|^{1/2}\exp\left[-nq(\hat{\theta})\right]}.$$

The reason for the improved convergence is that the error terms of $O(n^{-1})$ in the numerator and denominator are identical, canceling to leave the remaining higher-order term of $O(n^{-2})$. The proof is left as an exercise. For further discussion of asymptotic methods for integral approximation, see Tierny and Kadane (1986) and Kass et al. (1988, 1989).

Example B.2
Suppose that the parameter space can be partitioned as $\Theta = \Theta_1 \times \Theta_2$, where $\theta_2 \in \Theta_2$ is not of interest. We are interested in approximating the integral

$$E(g(\theta_1)|x) = \int_{\Theta_1}\int_{\Theta_2} g(\theta_1)p(\theta_1,\theta_2|x)d\theta_2 d\theta_1$$

$$= \frac{\int_{\Theta_1}\int_{\Theta_2}\exp\{\log[g(\theta_1)p(x|\theta_1,\theta_2)\pi(\theta_1,\theta_2)]\}d\theta_1 d\theta_2}{\int_{\Theta_1}\int_{\Theta_2}\exp\{\log[p(x|\theta_1,\theta_2)\pi(\theta_1,\theta_2)]\}d\theta_1 d\theta_2},$$

$$= \frac{\int_{\Theta}\exp\{\log[-nq^*(\theta)]\}d\theta}{\int_{\Theta}\exp\{\log[-nq(\theta)]\}d\theta}.$$

The Laplace approximation is

$$E[g(\theta)] \approx \frac{|\tilde{\Sigma}^*|^{1/2}\exp\left[-nq^*(\hat{\theta}^*)\right]}{|\tilde{\Sigma}|^{1/2}\exp\left[-nq(\hat{\theta})\right]},$$

where $\hat{\theta} = (\hat{\theta}_1,\hat{\theta}_2)$ maximizes $p(x|\theta)p(\theta)$, and $\hat{\theta}^* = (\hat{\theta}_1^*,\hat{\theta}_2^*)$ maximizes $g(\theta_1)p(x|\theta)p(\theta)$.

B.3   Monte Carlo Integration

A common criticism of Laplace's method is that it assumes 'smooth' functions. Suppose that we are interested in $p(\theta > 0|x)$ or $p(\delta_1 < g(\theta) < \delta_2|x)$, for a discrete parameter or a discontinuous transformation, respectively. Further complications can arise as the number of dimensions grows, or if the sample size is small. Obviously, these cases present problems, although they represent a large class of typical problems in bioinformatics. However, if we can obtain a sequence of values $\theta_1, \theta_2, \theta_3, \ldots$ sampled from the target distribution $p(\theta|x)$ then we can form approximations to integrals of interest to a desired degree of error. The following result provides a more general class of computational approximations that are very useful in a wide set of circumstances. Suppose that we are interested in evaluating

$$E\left[g(\theta)|x\right] = \int g(\theta)p(\theta|x)d\theta.$$

Then if $\theta_1, \ldots, \theta_n$ are i.i.d. samples from $p(\theta|x)$,

$$\frac{1}{n}\sum_{j=1}^{n} g(\theta_j) \rightarrow E\left[g(\theta)|x\right]$$

as $n \rightarrow \infty$ with probability 1 by the strong law of large numbers. This very general result does not require that $g(\theta)$ or $p(\theta|x)$ be smooth, or that $p(\theta|x)$ be unimodal. All we need is a large enough sample from $p(\theta|x)$, to achieve the desired accuracy. In practice, obtaining a sample from the posterior distribution can be direct or indirect, depending on whether or not a suitable scheme exists to perform sampling. If we can sample from $p(\theta|x)$ directly, then we are done, as we can rely on the strong law of large numbers to achieve the desired result. If not, a mechanism is needed to carry out indirect sampling, that is, sampling from a source other than $p(\theta|x)$, to yield a sequence from $p(\theta|x)$, or in more difficult situations, a sequence converging in distribution to $p(\theta|x)$.

Monte Carlo simulation is a widely and fundamentally accepted tool among Bayesians for estimation, inference, and prediction. Increasing the Monte Carlo size n of the sequence can lead to improved approximations, although many Bayesians make use of kernel methods to make approximations to posterior statements, once a reasonable sample from the posterior distribution is obtained. For example, the kernel estimate of $p(\theta|x)$ is

$$\hat{p}(\theta|x) = \frac{1}{nw_n}\sum_{j=1}^{n} K\left(\frac{\theta - \theta_j}{w_n}\right),$$

for the kernel function $K(\cdot)$, usually a normal or a symmetric unimodal distribution. The window width w_n controls the smoothness of the estimate; smaller windows provide better resolution, requiring a larger n. A useful result is that the window can be treated as a function of the Monte Carlo size n, satisfying $w_n \to 0$ and $n w_n \to \infty$ as $n \to \infty$ (Bernard and Silverman, 1986).

Example B.3

Suppose that $x_1, \ldots, x_n$ are i.i.d. $N(\mu, \sigma^2)$, $\pi(\mu, \sigma) \propto \sigma^{-1}$, and $\tau = \sigma^{-2}$, and we are interested in making posterior statements about the coefficient of variation, defined as $\psi(\mu, \sigma) = \sigma/\mu$. Recall that the posterior distribution of μ and $\tau = \sigma^{-2}$ is

$$\mu | \tau, x \sim N\left(\bar{x}, \frac{1}{n\tau}\right),$$

$$\tau | x \sim \text{Gamma}\left(\frac{n-1}{2}, \frac{(n-1)s^2}{2}\right),$$

for $s^2 = (n-1)^{-1} \sum_{i=1}^{n}(x_i - \bar{x})^2$. We generate samples in a sequence as follows:

1. Sample $\tau_j \sim \text{Gamma}\left(\frac{n-1}{2}, \frac{(n-1)s^2}{2}\right)$.

2. Sample $\mu_j | \tau_j \sim N\left(\bar{x}, \frac{1}{n\tau_j}\right)$.

3. Repeat steps 1 and 2 until a desired sequence $(\mu_1, \tau_1), \ldots, (\mu_n, \tau_n)$ of n observations is obtained.

We approximate $E(\psi|x)$ by

$$\hat{E}(\psi|x) = \frac{1}{n} \sum_{j=1}^{n} \psi_j = \frac{1}{n} \sum_{j=1}^{n} \tau_j / \mu_j.$$

Alternatively, for a $(1 - \alpha)$ highest posterior credible set for ψ, we can use the empirical $\alpha/2$ and $1 - \alpha/2$ quantiles of the sampled ψ_j. In this example, a sample from the posterior distribution can be obtained directly. We now turn to indirect sampling methods, which require much more consideration. These methods open up a richer set of computational tools, putting at our disposal a wide variety of Bayesian methods.

B.4   Importance Sampling

Suppose that we are interested in the posterior expectation of the transformation $g(\theta)$,

$$E\left[g(\theta)|x\right] = \frac{\int g(\theta) f(x|\theta) p(\theta) d\theta}{\int f(x|\theta) p(\theta) d\theta},$$

and, unlike the example above, no mechanism exists to sample directly from the posterior $p(\theta|x)$. If a density $q(\theta)$ exists that is similar to the posterior, and that one can sample directly, then Monte Carlo integration can be used, with the following modification. Since the sample θ_j, $j = 1, \ldots, n$, is from $q(\theta)$ rather than $p(\theta|x)$, then the average must be modified, with the appropriate weights. Consider the *importance weight function*

$$\omega(\theta) = \frac{f(x|\theta)p(\theta)}{q(\theta)},$$

the ratio of the unnormalized posterior to our density $q(\theta)$. Then

$$E\left[g(\theta)|x\right] = \frac{\int g(\theta)\omega(\theta)q(\theta)d\theta}{\int \omega(\theta)q(\theta)d\theta}$$

$$\approx \frac{\frac{1}{N}\sum_{j=1}^{N} g(\theta_j)\omega(\theta_j)}{\frac{1}{N}\sum_{j=1}^{N} \omega(\theta_j)},$$

where the θ_j, $j = 1, \ldots, N$, are sampled i.i.d. directly from $q(\theta)$. The density $q(\theta)$ is called the importance function.

There are many ways to choose the importance function $q(\theta)$ in practice. For all sets $A \in \theta$ such that the posterior $P(\theta \in A|x) > 0$, the envelope function must be greater than 0, $q(\theta \in A) > 0$, as well. That is to say, the more the importance function is similar to the posterior, the better the precision of the approximation. Not to be confused with prior specification, during posterior computation one may look at the data.

Importance sampling is appealing in its simplicity; all that is required to approximate integrals of (possibly) complex variable transformation with Monte Carlo integration is a sample from the importance function $q(\theta)$, and formation of the appropriate weights.

B.5   Rejection Sampling

For high-dimensional posterior parameter spaces, choosing an importance function that agrees well with the posterior can be a formidable task. Imagine that the posterior density is only known up to a normalizing constant,

$$p(\theta|x) \propto f(x|\theta)p(\theta),$$

in such a way that the normalizing constant cannot be evaluated. Posterior sampling can still be achieved, albeit with more sophisticated methods that we have examined so far.

Rejection sampling is a well-established technique used by Bayesians for its robustness and simplicity. Like importance sampling, we must specify a distribution $q(\theta)$ that is similar to the posterior that we can easily sample from, called the *envelope function*. Think of the envelope function as 'blanketing' the posterior, that is, we want a function $q(\theta)$ that is similar to and somewhat more dispersed that the posterior.

The steps of the rejection method are as follows. First, suppose that we have an envelope function $q(\theta)$ and a constant M such that

$$f(x|\theta)p(\theta) < Mq(\theta), \tag{B.1}$$

for all $\theta \in \Theta$, the parameter space. For each iteration j, we sample a *candidate* $\theta_j \sim q(\theta)$, and a uniform random variate $u_j \sim U(0, 1)$. If

$$u_j < \frac{f(x|\theta_j)p(\theta_j)}{Mq(\theta_j)}, \tag{B.2}$$

then we accept the candidate θ_j, and include it in our sample, and reject the candidate otherwise. The iterations are completed when the desired number of samples N are obtained. The accepted θ_j in our sample are distributed according to $p(\theta|x)$, although the number R of required iterations can vary from one sequence to the next. One measure of the efficiency of the algorithm is the acceptance ratio, the percentage of accepted candidates. In practice, we desire to generate a minimum expected number of iterations in order to obtain a sample of size N. Since the functions $f(x|\theta)$ and $p(\theta)$ are fixed, we must identify an envelope function $q(\theta)$ such that M is as small as possible, so that the maximum acceptance ratio is achieved. The random count of rejections proceeding a candidate's acceptance follows the well-known geometric distribution, with acceptance probability

$$P(\text{accept candidate}) = \frac{\int_{\Theta} f(x|\theta)p(\theta)d\theta}{M} \tag{B.3}$$

and expected value proportional to M. Hence, choosing M small furnishes a high acceptance rate; see Carlin and Luis (2000), Ripley (1987), and Devroye (1986).

Example B.4
In this example, the posterior is bimodal, a mixture of two Gaussian densities. The envelope function is specified as $N(0, 2^2)$, with $M = 3$, and achieved an acceptance ratio of 0.333, displayed in Figure B.1.

The following code was run in R to generate the sample:

```
f.star <- function(theta) dnorm(theta,-2,1)/3+2*dnorm(theta,1,.5)/3
f.env <- function(theta) dnorm(theta,0,2)
theta.seq <- seq(-8,8,length=1000)
N = 100000
M = 3
theta <- rnorm(N,0,2)
```

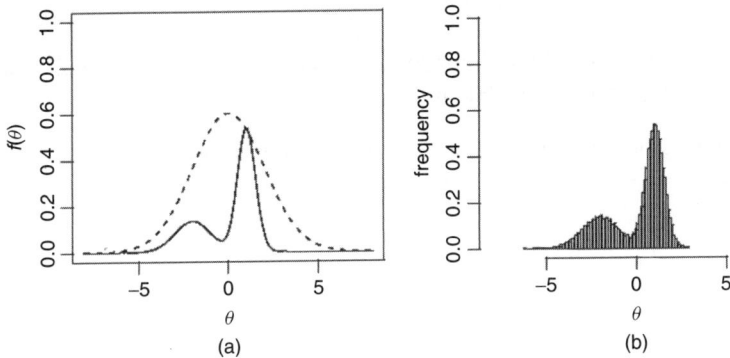

Figure B.1   (a) Target density (solid line) and envelope $q(\cdot) \cdot M$ (dashed line). (b) Histogram of sampled values.

```
u <- runif(N)
accept <- u < f.star(theta)/(M*f.env(theta))
theta <- theta[accept]
```

B.6   Gibbs Sampling

The Gibbs sampling algorithm is central to Bayesian applications and research. One attractive feature of the algorithm is its implementation of the mathematical principle of 'divide and conquer'. High-dimensional posterior parameter spaces can easily be sampled with Gibbs sampling, provided thatcertain robust conditions are met. Let us begin by defining a multivariate k-dimensional parameter space $\theta_1, \ldots, \theta_k \in \Theta_1 \times \ldots \times \Theta_k$, with joint posterior density denoted by $p_k(\theta_1, \ldots, \theta_k | x)$. We introduce the *posterior full conditional* distribution of θ_i, which is simply

$$p(\theta_i | \theta_1, \ldots, \theta_{i-1}, \theta_{i+1}, \ldots, \theta_k, x) \propto p(\theta_1, \ldots, \theta_i, \ldots, \theta_k | x), \qquad (B.4)$$

proportional to the posterior, holding all other parameter values as given, i.e. fixed. The posterior full conditional of θ_i is just the marginal posterior distribution of θ_i conditional on, or taking as given, $\theta_1, \ldots, \theta_{i-1}, \theta_{i+1}, \ldots, \theta_k$. The algorithm proceeds as follows:

1. Choose starting values $\{\theta_1^{(0)}, \ldots, \theta_k^{(0)}\}$.

2. Draw $\theta_1^{(1)}$ from its full conditional $p(\theta_1 | \theta_2^{(0)}, \ldots, \theta_k^{(0)})$.

3. Draw $\theta_2^{(1)}$ from its full conditional $p(\theta_2 | \theta_1^{(1)}, \theta_3^{(0)}, \ldots, \theta_k^{(0)})$ given $\theta_1^{(1)}$ drawn in the last step.

4. Draw $\theta_3^{(1)}$ from its full conditional $p(\theta_3 | \theta_1^{(1)}, \theta_2^{(1)}, \ldots, \theta_k^{(0)})$ given $\theta_1^{(1)}$ and $\theta_2^{(1)}$.

5. ...

6. Draw $\theta_k^{(1)}$ from its full conditional $p(\theta_k | \theta_1^{(1)}, \ldots, \theta_{k-1}^{(1)})$, conditional upon all parameter values drawn thus far.

One iteration of the Gibbs sampler is completed after we obtain one new sample for all k parameters, $\{\theta_1^{(1)}, \ldots, \theta_k^{(1)}\}$. The process is repeated R times to obtain a sample:

$$
\begin{array}{cccc}
\theta_1^{(0)}, & \theta_2^{(0)}, & \cdots & \theta_k^{(0)} \\
\theta_1^{(1)}, & \theta_2^{(1)}, & \cdots & \theta_k^{(1)} \\
\theta_1^{(2)}, & \theta_2^{(2)}, & \cdots & \theta_k^{(2)} \\
& \vdots & & \\
\theta_1^{(R)}, & \theta_2^{(R)}, & \cdots & \theta_k^{(R)}
\end{array}
\tag{B.5}
$$

from the joint posterior distribution of size R. Geman and Geman (1984) showed that the probability distribution of the sequence $\{\theta_1^{(t)}, \ldots, \theta_k^{(t)}\}$ converges to $p(\theta_1, \ldots, \theta_k | x)$ as $t \to \infty$, *exponentially*. Gibbs sampling was first introduced in the statistical literature by Gelfand and Smith (1990), and further discussed in Schervish and Carlin (1992).

If one is interested in the marginal density of a variable transformation of a subset of θ_is, the respective columns of (B.5) provide the sub-joint sample, i.e. ignoring the remaining columns. The marginal posterior of, for example, a function of $\psi(\theta_i)$ can be derived by extracting the ith column of (B.5), with marginal posterior estimated by

$$
\hat{p}(\psi(\theta_i)|x) = \frac{1}{R} \sum_{r=1}^{R} p(\psi(\theta_i)|\theta_1^{(r)}, \ldots, \theta_{i-1}^{(r)}, \theta_{i+1}^{(r)}, \ldots, \theta_k^{(r)}).
$$

Example B.5

Suppose $x_1, \ldots, x_n$ are i.i.d. $N(\mu, \sigma^2)$, $\tau = \sigma^{-2}$, and $\pi(\mu, \tau) \propto \tau^{-1}$. From previous examples, we obtain the full posterior conditionals as

$$
\pi(\mu|\tau, x) = N\left(\bar{x}, \frac{1}{n\tau}\right),
$$

$$
\pi(\tau|\mu, x) = \text{Gamma}\left(\frac{n}{2}, \frac{\sum(x_i - \mu)^2}{2}\right),
$$

The Gibbs sampler is as follows:

1. Generate starting values $(\mu^{(0)}, \tau^{(0)})$.

2. Generate $\mu^{(1)}$ from $\pi(\mu|\tau^{(0)}, x)$.

3. Generate $\tau^{(1)}$ from $\pi(\tau|\mu^{(1)}, x)$.

4. Repeat steps $2 - 3$ until we have R samples $(\mu^{(1)}, \tau^{(1)}), \ldots, (\mu^{(R)}, \tau^{(R)})$.

While Gibbs sampling provides great computational convenience for sampling high-dimensional posterior distributions, its utility is of course limited to situations where the full conditional distributions exist in known form and can easily be sampled. If the full conditionals are not easily sampled then we must resort to advanced methods. Notice that the full conditional expressions provide insight into the parameter dependence, and thus, the flow of information between the parameters, while updating at each Gibbs stage. The analytical convenience of working with these expressions cannot be overemphasized, as among other things, it can aid in choosing priors!

Example B.6
Suppose that $(x_{i1}, x_{i2})^T \sim N(\mu, \Sigma)$ are bivariate normal with μ unknown and

$$\Sigma = \begin{pmatrix} 1 & 0.90 \\ 0.90 & 1 \end{pmatrix}. \tag{B.6}$$

Let the prior for μ be flat, i.e. improper, and suppose that n observations are obtained. The Gibbs updating procedure can be performed by first specifying starting values, in this case $(0, 0)$. Next we can update the chain one dimension at a time for $r = 1, \ldots, 1000$ iterations:

1. $\mu_1^{(r)} \sim N(\bar{x}_1 + 0.90(\mu_2^{(r-1)} - \bar{x}_2), 1)$.

2. $\mu_2^{(r)} \sim N(\bar{x}_2 + 0.90(\mu_1^{(r)} - \bar{x}_1), 1)$.

Figure B.2 displays one realization of the resulting chains. The following code was run in R to generate the sample:

```
rho = .9
N = 1000
mu1 = mu2 = rep(0,N)
xbar1 = 1
xbar2 = 2
n = 25
sigma = sqrt((1-rho*rho)/n)
for(r in 2:N)–
mu1[r] = rnorm(1, xbar1+rho*(mu2[r-1]-xbar2), sigma)
mu2[r] = rnorm(1, xbar2+rho*(mu1[r]-xbar1), sigma)
"
```

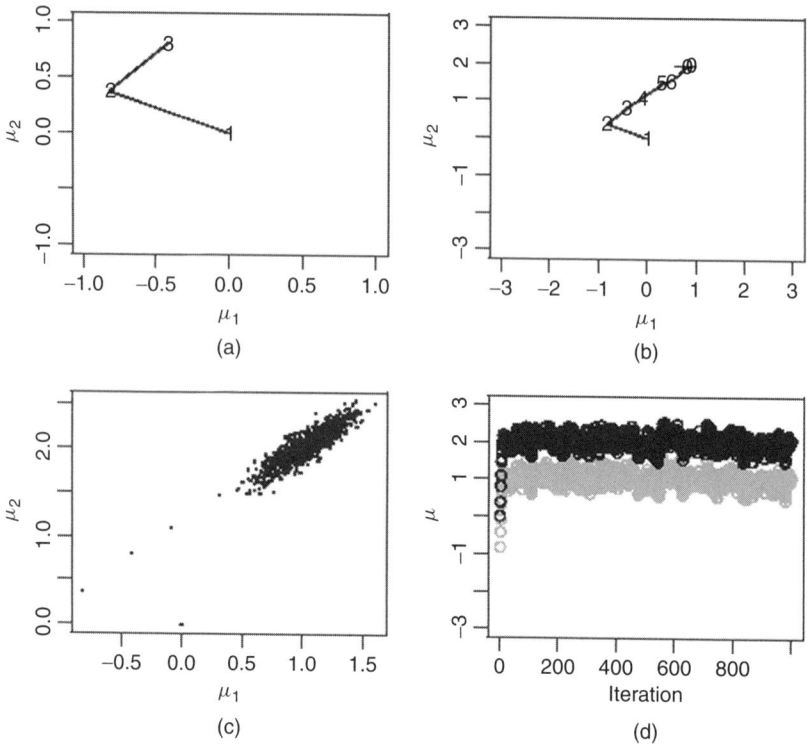

Figure B.2   (a) The first 3 observations in the chain. (b) The first 9 observations in the chain. (c) 1000 observations. (d) Trace plots, μ_1 (black open circles) and μ_2 (gray open circles).

Burn-in appears to be achieved after 10 or so observations, as the observations in the chain migrate away from (0,0). Note that choosing the starting values closer to regions of high posterior mass (i.e. in this case the sample mean $\bar{x}_1, \bar{x}_2$) works fine, and will speed up burn-in dramatically.

B.7   The Metropolis Algorithm and Metropolis–Hastings

Of the many difficulties in posterior computation, the most challenging hurdles are frequently those encountered in practice, calling for robust methods that can be applied in a variety of situations. Each indirect sampling scheme in the proceeding sections offers a solution, provided certain conditions are met, e.g. a suitable importance function exists for importance sampling, a convenient envelope function can be found for rejection sampling, or full conditionals are readily available for sampling with the Gibbs algorithm. As is often the case in practice, mild conditions can preclude the use of such methods, as even a simple tweak to a prior can have major implications for the efficiency of a computational tool. The Metropolis algorithm (Metropolis et al., 1953) is a general computational tool that is very popular among Bayesian for the flexibility it furnishes for indirect sampling of a posterior distribution. Hastings (1970) later proposed the algorithm for problems in statistics. The Metropolis and Metropolis–Hastings algorithms, discussed below, are *Markov chain Monte Carlo* methods, meaning that the methods generate a sequence of values moving from one state (i.e. the domain of the posterior in our case) to the next in a series of decisions, that in theory provide a chain of observations converging in distribution to the true target posterior.

Suppose that we desire a sample from the joint posterior distribution of $\theta = (\theta_1, \ldots, \theta_k)$, denoting the joint posterior density as $p_k(\theta)$. Let us introduce the *proposal density* $q(u, v)$, symmetric in its arguments, i.e. $q(u, v) = q(v, u)$. The proposal density provides candidate proposals to move between states, i.e. from v, the current state of the chain, to u, an *update* in the chain. In a single iteration t of the Metropolis algorithm, a candidate θ^* is generated as $\theta^* \sim q(\cdot, \theta^{(t-1)})$, conditional upon the current state of the chain, $\theta^{(t-1)}$. A decision is made to either accept the candidate θ^* and set $\theta^t = \theta^*$ or to reject the candidate, in which case $\theta^t = \theta^{t-1}$ and the chain does not 'move'. The candidate is accepted with probability equal to min$(r, 1)$, where

$$ r = \frac{p(v)}{p(u)} = \frac{f(x|\theta^*)\pi(\theta^*)}{L(x|\theta^{(t-1)})\pi(\theta^{(t-1)})}, $$

and rejected with probability $1 - $ min$(r, 1)$. A full implementation of the algorithm proceeds as with Gibbs sampling from a starting point $\theta^{(0)}$, e.g. the posterior mode or in some cases a perturbed estimate thereof, and is followed by the above iterations, accepting/rejecting candidates from the proposal, until the algorithm converges to the target posterior distribution, i.e., after the burn-in stage. Once it is determined that burn-in is achieved, many iterations are repeated until the desired number of samples are obtained. Unlike the rejection algorithm, the sequence generated by the Metropolis algorithm is a Markov chain and, as such, yields observations in a sequence that are dependent. A procedure called *thinning* is often employed to reduce the covariance between observations, for example, culling the sequence to retain only observations d iterations, or lags, apart.

While it may appear that the Metropolis algorithm is expensive to run, in high-dimensional settings, for example, remember that there is information embodied in the decision to move from the last state, which can inform the next decision, etc. Bayesians use all of the information! Under mild conditions the random sequence generated by the Metropolis algorithm converges in distribution to the target density, in our case the posterior, as $t \to \infty$ (Robert and Casella, 1999).

Example B.7

Suppose that $(x_{i1}, x_{i2})^T$ are i.i.d. bivariate $S(d, \mu, \Sigma)$, for $i = 1, \ldots, n$, with Σ known and $\mu = (\mu_1, \mu_2)^T$ unknown. We specify an improper flat prior for μ,

$$f(x|\mu)p(\mu) \propto \prod_{i=1}^{n} \left(1 + \frac{1}{d}(x_i - \mu)^T \Sigma^{-1}(x_i - \mu)\right)^{-2}.$$

For many problems with a continuous parameter space, a convenient choice for the proposal density is to generate a candidate,

$$\mu^* \sim N\left(\mu^{(t-1)}, R\right),$$

where $\mu^{(t-1)}$ is the current value of the chain. The proposal covariance R controls the variability in the 'jumps' of the proposal from the current state. Large jumps will tend to be accepted less often, using little information about the current state, whereas small jumps, while accepted at a higher rate, can yield chains with much more autocorrelation and take longer to explore the parameter space.

At iteration t, the odds ratio is computed,

$$r = \frac{\prod_{i=1}^{n}(1 + \frac{1}{d}(x_i - \mu^*)^T \Sigma^{-1}(x_i - \mu^*))^{-2}}{\prod_{i=1}^{n}(1 + \frac{1}{d}(x_i - \mu^{(t-1)})^T \Sigma^{-1}(x_i - \mu^{(t-1)}))^{-2}},$$

and compared with a uniform random variable $u \sim U(0, 1)$. If $u < \min(r, 1)$ then the candidate is accepted, i.e., $\mu^{(t)} = \mu^*$, otherwise we set $\mu^{(t)} = \mu^{(t-1)}$.

For continuous parameter settings, the most convenient choice for the proposal q is $N\left(\theta^{(t-1)}, \tilde{\Sigma}\right)$, where

$$\tilde{\Sigma} = -\left[\frac{\partial^2 \log[p^*(\theta|x)]}{\partial\theta\,\partial\theta'}\right]^{-1}\Big|_{\theta=\theta^{(t-1)}},$$

with $p^*(\theta|x) = f(x|\theta)p(\theta)$. This choice of q is easily sampled and clearly symmetric in v and $\theta^{(t-1)}$. In practice, choosing a good proposal variance is an acquired skill, requiring trial and error.

An important enhancement to the Metropolis algorithm was made by Hastings (1970), allowing for an asymmetric proposal density. Hastings generalized the odds ratio as

$$r = \frac{f(x|v)p(v)q(u, v)}{f(x|u)p(u)q(v, u)}.$$

For more discussion of MCMC methods, see Chib and Greenburg (1998) and Robert and Casella (1999).

Example B.8

Suppose that $(x_{i1}, x_{i2})^T \sim N((1, 2)^T)$, Σ), with covariance matrix

$$\Sigma = \begin{pmatrix} 1 & \rho \\ \rho & 1 \end{pmatrix} \tag{B.7}$$

and ρ unknown. Placing a flat prior on ρ, the log-posterior is

$$\log(p(\rho)) \propto -\frac{n}{2} \log |\Sigma| - \frac{1}{2} \sum_{i=1}^{n} \begin{pmatrix} x_{i1} & -\frac{1}{2} \\ x_{i2} & \end{pmatrix}^T \Sigma^{-1} \begin{pmatrix} x_{i1} & -\frac{1}{2} \\ x_{i2} & \end{pmatrix}. \tag{B.8}$$

Metropolis–Hastings updating is performed at iteration t, with beta proposal, letting $\alpha = (c - 2) * \rho^{(t-1)} + 1$ and $\beta = c - \alpha$, for $c = 25$. Why? The acceptance ratio in this example is

$$r = \exp \left[\log p(\rho^*|x) + \log q(\rho^{(t-1)}|x) - \log p(\rho^{(t-1)}|x) - \log q(\rho^{(t-1)}|x) \right].$$

Trace plots and histograms for 10 000 iterations are shown in Figure B.3. The acceptance rate, i.e. the rate at which the chain is updated to a new state, is 25%. The following code was run in R to generate the sample:

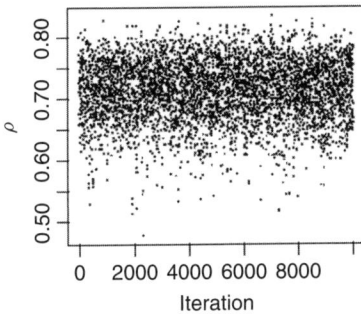
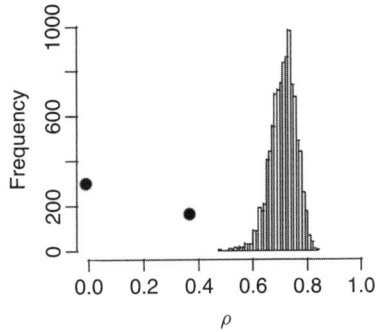

Figure B.3   (a) Trace plot for ρ. (b) Histogram of sampled ρs.

```
logpost <- function(rho)–
out = -(n/2) * log(1-rho*rho)
out = out - 1/(2*(1-rho*rho))*(sum(x1*x1) -2*rho*sum(x1*x2)
+ sum(x2*x2))
out
"

# generate bivariate normal data
N = 10000
```

```
R = 5000
true.rho = 0.70
V = matrix(c(1,true.rho,true.rho,1),2,2)
decomp = svd(V)
C = diag(sqrt(decomp$d))%*%decomp$u
n = 100
X <- matrix(rnorm(n*2),n,2)
Y <- X  rho = rep(0,R)
rho[1] = cor(Y)[1,2]
x1=Y[,1]
x2=Y[,2]
# *beta proposal
# candidate a* + b* = const
# a* = (const-2)*rho.old+1
# b* = const - a*
consts = c(1,4,10)
const = 3
for(r in 2:N)–
rho.old = rho[r-1]
a.star = (const-2)*rho.old+1
b.star = const - a.star
rho.star = rbeta(1,a.star,b.star)
a.star2 = (const-2)*rho.star+1
b.star2 = const - a.star
acc = logpost(rho.star) + log(dbeta(rho.old,a.star2,b.star2))
acc = acc - logpost(rho.old) - log(dbeta(rho.star,a.star,b.star))
if(log(runif(1)) < acc) rho[r] = rho.star else rho[r] = rho.old
”
```

B.8   Advanced Computational Methods

In practice, there are occasions when the methods outlined above do not provide the necessary tools to perform posterior sampling as desired. Much analytical and methodological effort has been put into overcoming such obstacles and providing improved efficiency for Monte Carlo applications. Fortunately, many advances have been made to combine the methods that we already discussed, or to apply them with simple modifications that can substantially improve performance. For example, recall that with Gibbs sampling an important condition is that the full conditionals can be sampled conveniently. If the full conditionals cannot be easily sampled, say because some collection of full conditionals have unknown form, one option is to perform the rejection algorithm *within* Gibbs sampling. Each full conditional is sampled in turn, until a full conditional of unknown form is encountered, $p(\theta_i|\theta_{[-i]}, x)$. The appropriate envelope function and constant M must be identified, and the rejection algorithm applied, until an observation is obtained from the full conditional. Rejection sampling within Gibbs was discussed in Gelfand and Smith (1990), Gilks and Wild (1992), Gelfand et al. (1990), and Wakefield et al. (1994). Metropolis within Gibbs has also been discussed (Gelfand, 2000).

B.8.1 Block MCMC

Thus far, we have discussed iterative schemes to update the sequence of parameter observations one dimension at a time. In some cases, the parameters can naturally be partitioned into blocks, or multivariate sets, that can be conveniently sampled in batches. Suppose, for example, that a multistage linear model is fitted, of the form

$$Y \sim N_n(\mu, \Sigma),$$

$$\mu \sim N_p(\Theta, \Omega),$$

$$\Theta \sim N_v(\Theta_0, \Omega_0),$$

where at the first stage the data Y are of dimension n, at the next stage μ is of dimension p, and at the deepest stage in the hierarchy Θ is of dimension v. For simplicity, let us take Θ_0, Σ, Ω and Ω_0 as known, and let parameter vectors μ and Θ be unknown. The model can be expressed as a system of linear equations

$$Y = \mu + E,$$

$$\mu = \Theta + W,$$

$$\Theta = \Theta_0 + W_0,$$

where $E \sim N_n(0, \Sigma)$, $W \sim N_n(0, \Omega)$, and $W_0 \sim N_n(0, \Omega_0)$. This system of equations can conveniently be reexpressed as

$$Y = \mu + 0 + E,$$

$$0 = -\mu + \Theta + W,$$

$$-\Theta_0 = 0 + \Theta + W_0,$$

so that a new multivariate normal response can be defined as

$$\tilde{Y} = \begin{pmatrix} Y \\ 0 \\ -\Theta_0 \end{pmatrix}$$

with mean

$$\tilde{\mu} = \begin{pmatrix} \mu \\ -\mu + \Theta \\ \Theta \end{pmatrix} = \begin{pmatrix} 1 & 0 \\ -1 & 1 \\ 0 & 1 \end{pmatrix} \begin{pmatrix} \mu \\ \Theta \end{pmatrix}$$

and covariance $\tilde{\Sigma}$, of block diagonal structure, with Σ, Ω, and Ω_0 along the diagonal. Sampling techniques, discussed in Hodges (1998) can be used to obtain posterior samples of the unknown parameter vectors. The above reexpression can be applied to the case where there are many more stages. When p and v are very large, or if there are many stages to consider, it may be more practical to sample the parameters in blocks, i.e. sample the vectors $\mu|\Theta, \cdot$ and $\Theta|\mu, \cdot$ in turn. This is called block sampling, and can be applied with the methods discussed above (Wilkinson and Yeung, 2002). It is advantageous, since, if the vector of parameters θ has known full conditional distribution, it can be sampled simultaneously, rather than sampling each individual dimension θ_1, followed by θ_2, etc.

B.8.2   Truncated Posterior Spaces

Suppose, for example, that, rather than the full parameter space, it is known that the p-dimensional mean of the multivariate Gaussian response Y lies in the continuous line between vectors a and b, i.e. the posterior parameter space lies in a hypercube with hyperplane manifolds specified by the elements in the vectors a and b. A normal prior would appear advantageous, if the constraints were not imposed. Suppose that a normal prior is specified as $p(\mu) \propto N(0, R) \times I(a, b)$ where the identity function is one if $\mu \in (a, b)$ and zero otherwise. Conveniently, we sample the elements of μ one at a time, as μ_1, $\mu_2|\mu_1$, $\mu_3|\mu_2$, μ_1, etc., where at each stage, the density of the parameter μ_j is specified as the appropriate conditional normal truncated to be in (a_j, b_j). This algorithm will produce multivariate observations, within the appropriate bounds.

A more challenging situation arises when the parameters lie in an amorphous set of nonlinear manifolds. The above algorithm can still be applied, taking care to evaluate the appropriate bounds at each stage, e.g. for μ_j given the parameter values $\mu_1, \ldots, \mu_{j-1}$ already sampled. For example, suppose that the parameter space is two-dimensional and bounded within the unit sphere. Then μ_1 is bounded to lie in $(-1, 1)$. Once μ_1 is sampled then μ_2 is bounded to be between $\pm\sqrt{1 - \mu_1^2}$.

B.8.3   Latent Variables and the Auto-Probit Model

Suppose that random binary responses $Y_1, Y_2, \ldots, Y_n$ exhibit dependence, defined through the $n \times n$ adjacency matrix A, i.e. the (i, j)th entry of A equals 1 if Y_i and Y_j are dependent, and zero otherwise. There are many ways to define dependence, but for now, let us take the dependence between pairwise neighbors to be defined globally. In the Bayesian probit modeling framework, the marginal binary responses can be fitted by introducing a latent normal variable Z such that

$$P(Y_i = 1 | Z_i > 0) = 1,$$
$$Z_i \sim N(\theta, \sigma^2),$$

that is, the conditional distribution of $p(Z_i | Y_i = 1) \propto N(\theta, \sigma^2) \times I(Z_i > 0)$ and $p(Z_i | Y_i = 0) \propto N(\theta, \sigma^2) \times I(Z_i < > 0)$ (Chib and Greenberg, 1998). Posterior updating for θ and σ^2 are made conditional upon the true labels, i.e. conditional upon the latent Zs during updating, while the Zs are updated, given Y, i.e. truncated, conditional upon sampled θ and σ^2.

For multivariate binary response data, the prior density of the vector Z is specified as

$$Z \sim N(\theta, (I - \rho A)^{-1}),$$

multivariate Gaussian, for the unknown mean θ, scalar ρ, and known adjacency matrix A. The full conditional prior, setting $\theta = 0$ without loss of generality, for Z_i is $N(\rho \sum_{i \sim j} z_j, 1)$, with the notational convention that the ith and jth entries are neighbors if $i \sim j$. The full conditional posterior for Z_i is updated with sign restriction, given Y_i, and ρ. Placing a flat prior on ρ, the Metropolis algorithm proceeds subject to the condition that $(I - \rho A)$ is *positive definite*; see Weir and Pettit (2000) for details on model fitting and imputation of missing data. Intuitively, this is a nearest-neighbor model, with the vote of neighbors depending on ρ, through the conditional mean. Thus, the unknown parameter ρ tells us the strength of borrowing between neighbors.

B.8.4 Bayesian Simultaneous Credible Envelopes

A serious challenge facing statisticians is estimation of variation and credible (or confidence) bounds. For a frequentist, this entails derivation of the appropriate asymptotic frequency confidence bounds, if the exact bounds cannot be found, for each new fitted model. This challenge is compounded by many orders of magnitude for multivariate estimates, e.g. function data analysis, or spatial analysis. In functional data analysis one may desire a set of plausible outcomes, given specified covariate values, or a simultaneous set of outcomes across the functional range, given the complete domain of covariate values. The theoretical challenges for deriving simultaneous functional data confidence sets are well known. For example, with nonlinear regression, the asymptotic approximations depend on the smoothness of the function, and are known to exhibit problems at discontinuities and loci with sharp second derivative.

For a Bayesian, there are no obstacles to credible set estimation, once a collection of observations are obtained from the posterior. Suppose that the response y is not a single observation, but rather a function of t for $t = 1, \ldots, T$. Then the response is a curve, and predictions nearby, say for t and $t + 1$, can be dependent.

Let us assume that the response is

$$y(t; \theta) = f(t; \theta) + \epsilon(t),$$

for $t = 1, \ldots, T$, where $f(t; \theta)$ is a function, not necessarily smooth, of t and θ, with $\epsilon(t)$ i.i.d. $N(0, \sigma^2)$. Suppose further that a proper prior is specified on θ and σ^2 as $p(\theta, \sigma^2)$, and given observations $y(1), \ldots, y(T)$, the posterior density is obtained.

Generating $r = 1, \ldots, R$ observations from the posterior, a $(1 - \alpha)100\%$ simultaneous credible envelope for the function $f(\theta)$ is found numerically, by sorting the posterior evaluated at each $\theta^{(r)}$, i.e. $p(\theta^{(r)}|y)$, ranking these values from smallest to largest, and choosing curves $f(1, \ldots, T; \theta^{(r)}) : p(\theta^r|y) \in (p(\theta^r|y)^{\alpha/2}, p(\theta^r|y)^{1-\alpha/2})$, i.e., such that the posterior evaluated at $\theta^{(r)}$ is within its numerical $(1 - \alpha)100\%$ bounds. We call this a Bayesian credible envelope (BCE) for the function $f(\theta)$. As the number of MCMC samples increases, the BCE bounds will resolve with increasing clarity. Determining simultaneous theoretical confidence envelopes for frequentist estimators of θ is much more involved, requiring assumptions about f and the underlying random error distribution for ϵ. Note that the above technique for deriving BCEs is robust to normality and independence. Readers interested in frequentist function data estimation are referred to Genovese and Wasserman (2004).

B.8.5 Proposal Updating

One of the most controversial and exciting areas of MCMC thinking involves continuous updating of the proposal $q(u, v)$. Recall that a good proposal density can make a big difference, improving the efficiency and time to generate samples from the posterior. Here we address the topic of proposal updating. Can one learn to improve a proposal continuously? It would appear that information in the Monte Carlo chain should help in generating better candidate proposals, based on acceptance ratios, and other criteria. In general, there is no general formula for updating proposal densities, and in practice many would doubt that such a routine would ever converge. There are exceptions, and we turn to an important one for multivariate posterior generation.

Consider multivariate observations $Y_i|\theta$, of dimension p, and suppose that the joint prior for the v-dimensional parameter θ is specified as $p(\theta) \neq \Pi_{j=1}^{v} p(\theta_j)$, i.e. dependence. Ignoring posterior parameter dependence while updating can be inefficient. Imagine we are trying to fill a mass stretched over an affine hyperplane, in some set of subdimension v. Choosing a joint proposal that takes into account dependence can avoid jumps outside of the affine plane, and save time, i.e. improve acceptance rates. In practice, though, determining the dependence can be tricky. The following algorithm has been suggested for just such a situation. Suppose that the parameter space is continuous.

1. Transform variables to lie on the real line, $\Theta \rightarrow \Upsilon$, and to stabilize variance.

2. Run Metropolis–Hastings and derive a preliminary estimate of the covariance of $\upsilon \in \Upsilon$.

3. Apply MCMC with a normal proposal, using the covariance estimated from the sample from step 2.

Comparing the acceptance rates between steps 2 and 3, there can be remarkable improvements. See (Carlin and Louis, 2000).

B.9   Posterior Convergence Diagnostics

Recall that the Gibbs and Metropolis algorithms create Markov chains, which under suitable conditions converge to the appropriate target (in our case posterior) distribution. In MCMC theory, the ultimate goal is to establish that the chain will reach its appropriate *stationary* distribution. Unfortunately, there are no general results to guarantee the number of iterations required for convergence, or even suitable evidence that by iteration N in the sequence, to within some tolerance, the MCMC chain converges.

There are many diagnostic methods available for monitoring MCMC convergence, many of which follow from two basic strategies:

1. Run one long chain, and compare sample statistics of interest, e.g. the mean or range of quantiles, along fixed windows.

2. Run many chains, from different starting values, and compare sample statistics of interest between them, i.e. infer significant variation between the chains, relative to the variation within each chain.

There is no consensus on how to run MCMC chains or monitor convergence. Nor is it likely that a general consensus will develop. There are many difficulties to consider in achieving convergence, such as the length of the chains, the modality and covariance of the posterior, and regions of the support over which much of the posterior is relatively flat. In the case where the posterior distribution lies largely along an affine hyperplane of the full dimension space, or the parameter space is distributed nonuniformly, i.e. with discrete jumps, between the dimensions, one can expect to encounter chains that mix slowly with conventional updating tools. In some cases, a one-to-one transformation, if it exists, among the parameters, can reduce drag, although in practice these problems can be very difficult to deal with, and to monitor convergence. Other issues to consider with MCMC convergence include models that are overparameterized, or full conditionals that do not admit a legitimate joint probability distribution. For instance, it is trivial to assign conditional distributions $(Z_i|Z_j)$ specified in such as way that the covariance is asymmetric, a serious problem for statisticians and probabilists, although less so for others using MCMC methods. We can disregard such cases here, as we are interested in legitimate joint probability distribution, but beware that there are many who insist that such models can perform well.

In theory, methods for monitoring convergence, including both graphical and numeric, can be shown to work reasonably well for low-dimensional parameter spaces. In practice, especially with high-dimensional parameter spaces, the reality is that true convergence can never really be known. In fact, many approaches to inferring convergence serve more to ease the mind of the analyst, providing psychological comfort, rather than direct proof, which cannot be had, that a chain has converged.

Suppose that one is interested in generating a sequence $\theta_1, \theta_2, \ldots$ from a distribution $p(\theta|x)$. In Monte Carlo estimation the error can be partitioned into two sources, that which arises from random moves between states, and that which arises during the initial period during which the chain is updating, and not yet converged. Suppose that $\int |p_t(\theta|x) - p_{t-1}(\theta|x)|d\theta \to 0$ as $t \to \infty$, and also that we can approximate $p_t(\theta|x)$ by $\hat{p}_t(\theta|x)$, i.e. with noise. By the triangle inequality,

$$\int |p_t(\theta|x) - p_{t-1}(\theta|x)|d\theta \leq \int |\hat{p}_t(\theta) - p_t(\theta)|d\theta$$

$$+ \int |p_t(\theta|x) - p(\theta|x)|d\theta.$$

Both terms on the right must go to zero as $t \to \infty$ for the sequence $\hat{p}_t(x)$ to converge to $p(x)$. Diagnostic methods cannot 'prove' overall convergence. Diagnostic monitoring and checking are concerned with divergence between $\hat{p}_t(\theta|x)$ and $\hat{p}_{t+k}(\theta|x)$ within a chain, or divergence between m different chains $\hat{p}_t^1(\theta|x), \hat{p}_t^2(\theta|x), \ldots, \hat{p}_t^m(\theta|x)$, as a function of N, the number of iterations. Unusual behavior in the chains is certainly a sign of problems, but lack of such clues is not evidence to conclude convergence, especially when a posterior is high-dimensional, or only known up to a normalizing constant. The chains may be near convergence, although if little is known about the full posterior, it may be that a seemingly well-behaved chain or collection of chains, for example, have not toured the entire parameter space.

These issues, and many like them, prevent us from ever determining with complete certainty whether or not a Monte Carlo chain has converged. In the sections that follow, we turn to traditional approaches for monitoring MCMC convergence, illustrating the affect of the proposal and offer some guidance for Bayesian applications with high-throughput data.

B.10   MCMC Convergence and the Proposal

Thus far, we learned that the proposal density selection is very important to achieve adequate acceptance rates. The proposal density is of course subjective.

Admittedly, choosing a good proposal is a black art to some, but we consider it an acquired skill that Bayesians must master. There is much practical advice in the literature, and we do not attempt to review it all here. Many good examples can be found in other texts that offer a complete introduction to training MCMC chains to achieve convergence.

Recall that, in the Metropolis algorithm, a candidate θ^* proposal is accepted with probability equal to $\min(r, 1)$, where

$$ r = \frac{f(x|\theta^*)\pi(\theta^*)}{L(x|\theta^{(t-1)})\pi(\theta^{(t-1)})}, $$

setting $\theta^{(t)} = \theta^*$, and rejected with probability $1 - \min(r, 1)$, in which case we set $\theta^{(t)} = \theta^{(t-1)}$. Consider indexing a symmetric proposal $q_v(\theta)$ by the variance v of the proposal density. Large values of v will tend to proposal candidates θ^* further from the current state of the chain $\theta^{(t-1)}$, which can result in lower acceptance rates. The last candidate to be accepted equals $\theta^{(t-1)}$, i.e. offering information about the relative probability of moving near this state. Moves that are too far from accepted candidates can suggest transitions to regions of minimal posterior mass. On the other hand, small values of v tend to yield higher acceptance rates, while providing very short moves between states. Such chains can take a very long time to explore the entire parameter space, while suffering from severe autocorrelation.

As a general rule of thumb, acceptance rates between 30% and 40% are considered reasonable. Acceptance rates lower than this tell us that the chain is not mixing well, an inefficient use of our resources. Acceptance rates higher than 40% might suggest high autocorrelation in the chains, in which case the observation in the sequence cannot be taken as a random sample from the posterior.

Example B.9
We return to Example B.8, with bivariate normal data and unknown correlation ρ. In this example, no mechanism exists to sample ρ directly. We are interested in the properties of the Monte Carlo chains, as a function of starting values and the variance of the proposal.

In Figure B.4 we display 100 iterations of four chains of the Metropolis algorithm constraining the proposal $\alpha^* + \beta^* = 100$, corresponding to a relatively moderate proposal variance. While the chains began at different starting values, these tend to be converging in distribution over similar support of the parameter space. Figure B.5 shows the results of four different chains, each with different proposal variance, as measured by $\alpha^* + \beta^* = c$, with larger values of c corresponding to smaller proposal variances. Figure B.5(a) has very high acceptance ratios, although the chain appears to be wandering about, not really depicting a random sample from the posterior. Figure B.5(d) shows a more moderate temporal trend and less autocorrelation. The autocorrelation functions for each chain in Figure B.5 are shown in Figure B.6. High autocorrelation can be reduced by

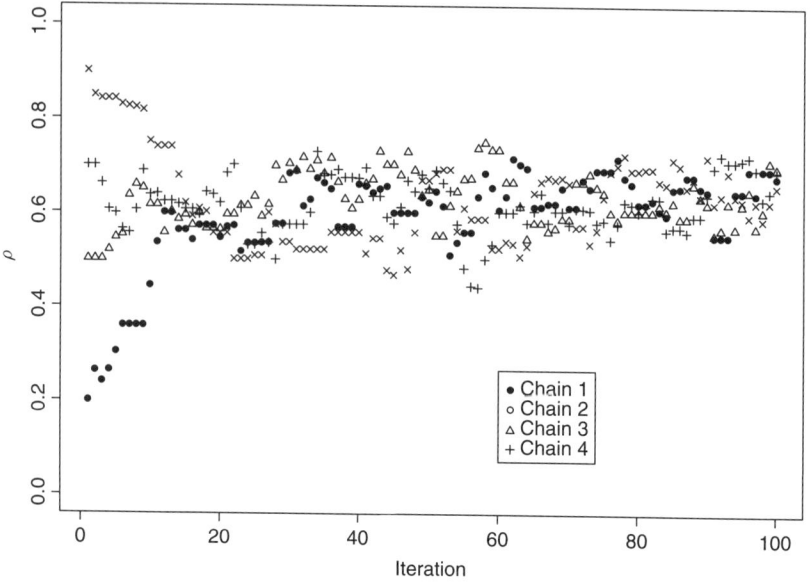

Figure B.4   MCMC chain initiated from four different starting values: 0.20, 0.50, 0.70, and 0.90.

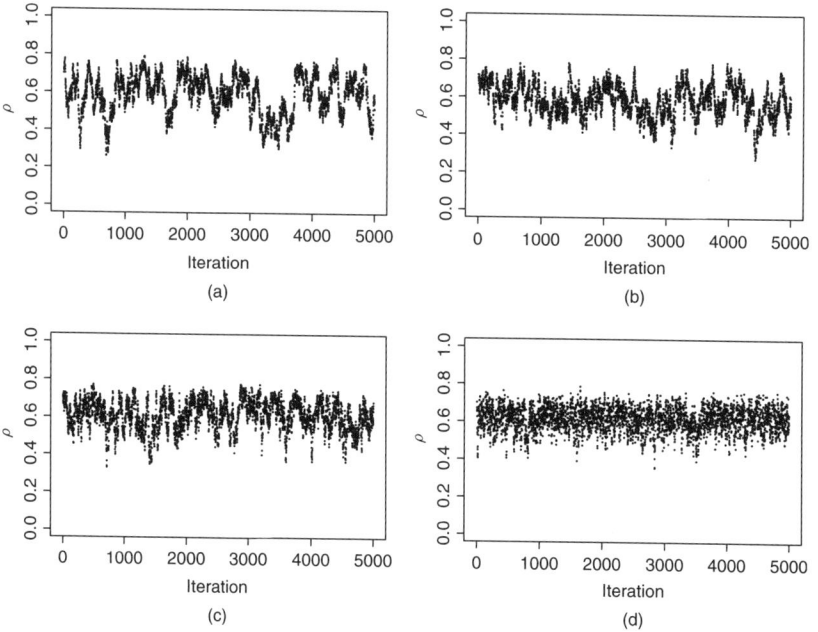

Figure B.5   MCMC chains each with different proposals for $\alpha^* + \beta^*$: (a) 700, (b) 500, (c) 300, and (d) 100.

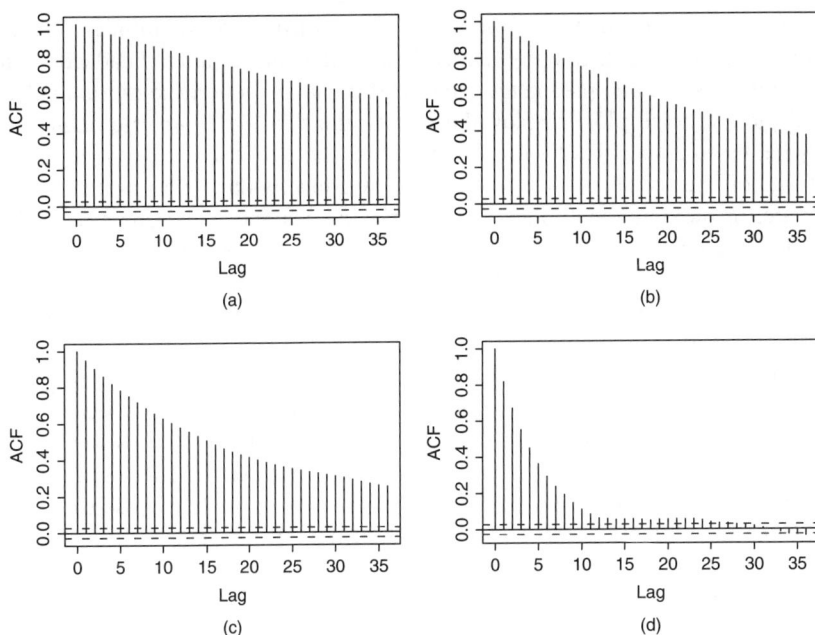

Figure B.6 Autocorrelation functions for the chains in Figure B.5.

a process called thinning, where only observations d lags apart in the sequence are retained.

The results in the above example beg the question, has the chain converged? Imagine if we had 10 000 such parameters. How could we monitor the convergence of all of them? How many chains should we run, and for how many iterations? How should we choose the starting values? In many situations, the answers to these questions are limited by available resources, although there are not necessarily right or wrong ways to perform MCMC. We turn to some general diagnostic results in the next section.

B.10.1   Graphical Checks for MCMC Methods

Ironically, most MCMC diagnostic inference involves tools from our fundamental frequentist arsenal to compare the functions of MCMC chains, along blocks of the sequence, or across chains. Diagnostic plots can be very helpful for getting a rough estimate of where burn-in is achieved. The trace plots shown in Figures B.4–B.5 reveal that much longer chains must be realized before one should begin to accept the observations as coming from the target, or stationary, distribution. These plots reveal trends in the chain that are vitally important, such as drift, and how well the chains are *mixing*, that is, exploring the parameter space, something that statistics cannot tell us. For example, the analyst hardly ever wishes to observe a scale parameter $\sigma^{(t)}$ for $t = 1, \ldots$ drifting off to infinity, or zero, although diagnosing such a problem early on can avoid countless headaches later, as a single drift in one parameter can set off a chain reaction (no pun intended!).

Typically, trace plots display the observations in the chain, although it can also be useful, specifically for functions of the sequence, to plot a running statistic of the sequence, for a specified window width, such as the moving average, or moving percentiles, e.g. the moving 5th and 95th percentiles. Much thought has gone into considerations of chain length, but in practice this is a function of the starting values, the proposal, and many other factors, that in general complicate the choice of chain length. One proposal is to monitor the ratio of the mean of a running statistic across chains, relative to the pooled running statistic from all chains. This is the approach implemented in the WinBugs software (http://www.mrc.bsu.cam.ac.uk/bugs) to monitor chains graphically using the range of the posterior 80th central percentiles. Obviously this method has limitations, as it is a graphical tool, and a univariate one at that. Other helpful figures include kernel density plots and boxplots which can help to detect outliers and other signs that a chain has drifted. Starting MCMC chains from different starting values is a good idea, to check that the chains are mixing properly, and moving to the same support.

B.10.2   Convergence Statistics

When has a Monte Carlo Chain converged to the stationary distribution? One of the most cited and widely available statistics for comparing chains was proposed by Gelman and Rubin (1992). The method is simple and quite intuitive. Suppose that we are interested in a parameter θ. Many chains can be run, and we can compare the overall variance between chains to the variance within the chains, much like the standard approaches to frequentist model selection and inference, i.e. the F-test procedure. The steps are as follows:

1. Choose K, the number of parallel chains to run, each from different starting values. It is important that the chains be initially *overdispersed* with respect to the true posterior, in order, *inter alia*, to minimize the chance of a chain getting stuck in a region of high mass, that it cannot get out of early on.

2. Each chain is run for $2N$ iterations.

3. We compute the *scaled reduction factor*,

$$\sqrt{\hat{R}} = \sqrt{\frac{N-1}{N} + \frac{m+1}{mN}\left(\frac{B}{W}\right)\left(\frac{df}{df-2}\right)},$$

where B/N is *variance between* the means from the m parallel chains, W is the average of the m *within-chain variances*, and df is the degrees of freedom of an approximating t density to the posterior distribution. As the length of the chains $N \to \infty$ then $\hat{R} \to 1$ evidence of good convergence.

In multivariate settings monitoring convergence is much more difficult. Rather than monitor each parameter, one could monitor, for instance, the logarithm of the posterior evaluated over the full parameter set.

Geyer (1992) proposed a different approach, running one long chain to monitor the variance of

$$\hat{\theta}_n = \frac{1}{N} \sum_{b=1}^{n} f(\theta^{(b)})$$

for N iterations. By the central limit theorem, under certain regularity conditions, we have that

$$\sqrt{n} \frac{(\hat{\theta}_n - E\theta)}{\sigma} \to N(0, 1).$$

This suggests that as the chain length increases, we should observe the sample average of the sequence converging in distribution to normality. A complication with this, of course, is that the unbiased estimation of σ^2 is unlikely in the presence of autocorrelation. Let us consider dividing the chain into m chains of equal size, that are approximately uncorrelated (this should be verified). We can estimate σ^2/n by

$$\hat{\sigma}^2 = \frac{1}{m(m-1)} \sum_{h=1}^{m} (\hat{\theta}_{n/m}^h - \hat{\theta}_n)^2$$

where $\hat{\theta}_{n/m}^h$, for $h = 1, \ldots, m$, is the sample mean in the hth chain. A 95% confidence interval for $E(\theta)$ is

$$\hat{\theta} \pm t_{m-1,.025} \sqrt{\hat{\sigma}^2}$$

Raftery and Lewis propose estimating σ^2 using the technique of thinning, keeping every dth sample after burn-in. We agree with Raftery and Lewis that a very sensible suggestion is to choose N and d large enough to estimate quantities such as $P(\theta < c|y)$ to within tolerable limits, of course taking the goals of the analysis into account.

B.10.3   MCMC in High-throughput Analysis

The illustrations that we have provided thus far serve to illustrate the complexity of monitoring MCMC convergence. As demonstrated in Example B.9, monitoring one-dimensional MCMC is complicated enough. Now imagine repeating the process several thousand times. This is absurd and will not be done in practice. Nevertheless, we provide some reasonable advice for Bayesians performing high-throughput data analysis. First of all, if you are considering a Bayesian analysis, try to avoid MCMC methods such as Metropolis altogether. If you insist on a sophisticated analysis, consider Monte Carlo integration of other forms, such as importance sampling. Or if feasible, attempt to rely on the Bayesian central limit theorem; treating the likelihood as approximately normal-gamma can greatly reduce the computational complexity. If it is not possible to further reduce the computational sophistication, at least attempt to pose the computation algorithm in a Gibbs framework. Other examples of simplifications that can greatly ease computation costs include reasonable approximations to full conditionals, approximations of one-to-one parameter transformation, and marginalizing over nuisance parameters that are difficult to sample indirectly. These devices can greatly alleviate the computational difficulty, freeing precious time for other stages of the analysis.

Parallel processing is making it possible to perform Monte Carlo sampling at rates that were once unimaginable. As a practical step, the output of such procedures must be carefully followed for signs of that the chains are healthy, such as trends in dispersion parameters or mixture weights. If the dimensionality is to high to monitor the chains one by one, consider spot checking, moving averages, or Gelman-Rubin statistic or Raftery–Lewis statistics can be helpful. Remember, the goal is to diagnose problems with convergence early on, since proving convergence is unlikely. Readers interested in further exploration should consult Cowles and Carlin (1996), Schervish and Carlin (1992), and Robert and Casella (1999).

B.11   Summary

There is no right or wrong way to utilize the Bayesian computational tools discussed in this appendix, although some methods are less efficient than others, or lead to poor approximations of posterior probabilities of events of interest. The methods outlined are very powerful, but should be used thoughtfully, considering such matters as acceptance ratios, the dimensionality of the data, and alternate methods. For example, the Laplace method, and extensions, provide a useful tool for making posterior statements, in situations where the posterior can be approximated well and the parameter space is continuous. In many situations, such as mixture modeling or hierarchical modeling, better alternative methods involve updating procedures with indirect sampling such as Gibbs sampling and Metropolis–Hastings, see Chapter 7. The basic outline provided in this chapter is an introduction and a guide, rather than a comprehensive overview. Posterior computation is an art, and we advise those interested to practice, starting with simple examples.

There are no easy solutions or recipes for monitoring chains, and it is best in practice to keep the process under control. Recall that if the goal is to generate samples from $p(\theta|y)$ with Monte Carlo methods then we must at least consider:

(a) the dimensionality of the parameter space;

(b) whether the full posterior conditionals lead to a legitimate joint posterior distribution;

(c) the length of the chains;

(d) when burn-in is achieved;

(e) autocorrelation in the chains;

(f) drift in the chains, i.e. to local modes.

For the novice, we suggest practicing with simulated examples before attempting advanced modeling. For those considering advanced modeling, keep in mind that monitoring convergence is a difficult, and at some times frustrating, task, although the rewards of mastering the art of performing MCMC sampling are well worth the effort.

References

Akaike, H. (1974) A new look at statistical model identification, *IEEE Transactions on Automatic Control*, 19, 716–723.

Albert, J. and Chib, S. (1993) Bayesian analysis of binary and polychotomous response data. *Journal of the American Statistical Association*, 88, 669–679.

Alizadeh, A., Eisen, M., Davis, R.E., Ma, C., Sabet, H., Tran, T., Powell, J.I., Yang, L., Marti, G.E., Moore, D.T., Hudson, J.R. Jr, Chan, W.C., Greiner, T., Weisenburger, D., Armitage, J.O., Lossos, I., Levy, R., Botstein, D., Brown, P.O., and Staudt, L.M. (1999) The lymphochip: a specialized cDNA microarray for the genomic-scale analysis of gene expression in normal and malignant lymphocytes. *Cold Spring Harbor Symp. Quant. Biol.*, 64, 71–78.

Alizadeh, A.A., Ross, D.T., Perou, C.M., and van de Rijn M. (2001) Towards a novel classification of human malignancies based on gene expression patterns. *J. Pathol.*, 195(1), 41–52.

Allison, D.B., Gadbury, G.L., Heo, M., Fernández, J.R., Lee, C.-K., Prolla, T.A., and Weindruch, R. (2002) A mixture model approach for the analysis of microarray gene expression data. *Computational Statistics & Data Analysis*, 39(1), 1–20.

Allocco, D.J., Kohane, I.S., and Butte, A.J. (2004) Quantifying the relationship between co-expression, co-regulation and gene function, *BMC Bioinformatics*, 5.

Alon, U., Barkai N., Notterman, D.A., Gish, K., Ybarra, S., Mack, D., and Levine, A.J. (1999) Broad patterns of gene expression revealed by clustering analysis of tumor and normal colon tissues probed by oligonucleotide arrays. *Proc. Natl. Acad. Sci. USA*, 96(12), 6745–6750.

Ambroise, C., and McLachlan G.J. (2002) Selection bias in gene extraction on the basis of microarray gene-expression data. *Proc. Natl. Acad. Sci. USA*, 99(10), 6562–6566.

Androulakis, I., Yang, E., and Almon, R. (2007) Analysis of time-series gene expression data: methods, challenges and opportunities. *Annual Review of Biomedical Engineering*, 9, 205–228.

Antoniak, C.E. (1974). Mixtures of Dirichlet processes with applications to Bayesian nonparametric problems. *Annals of Statistics*, 2, 1152–1174.

Arratia, R., Barbour, A.D. and Tavaré, S. (1992) Poisson process approximations for the Ewens sampling formula. *Annals of Applied Probability*, 2, 519–535.

Arteaga-Salas, J.M., Zuzan, H., Langdon, W.B., Upton, G.J., and Harrison, A.P. (2008) An overview of image-processing methods for Affymetrix GeneChips. *Briefings in Bioinformatics*, 9(1), 25–33.

Aronszajn, N. (1950) Theory of reproducing kernels. *Transactions of the American Mathematical Society*, 68, 337–404.

Ashburner, M., Ball, C.A., Blake, J.A., Botstein, D., Butler, H., Cherry, J.M., Davis, A.P., Dolinski, K., Dwight, S.S., Eppig, J.T., Harris, M.A., Hill, D.P., Issel-Tarver, L., Kasarskis, A., Lewis, S., Matese, J.C., Richardson, J.E., Ringwald, M., Rubin, G.M., and Sherlock, G. (2000) Gene ontology: tool for the unification of biology. The Gene Ontology Consortium. *Nat. Genet.*, 25, 25–29.

Bae, K., and Mallick, B.M. (2004) Gene selection using a two-level hierarchical Bayesian model. *Bioinformatics*. (18): 3423–3430.

Baggerly, K.A., Coombes, K.R., Hess, K.R., Stivers, D.N., Abruzzo, L.V., and Zhang, W. (2001) Identifying differentially expressed genes in cDNA microarray experiments. *J. Comput. Biol.*, 8(6), 639–659.

Baladandayuthapani, V., Holmes, C.C., Mallick, B.K., and Carroll, R.J. (2006). Modeling nonlinear gene interactions using Bayesian MARS. In K.A. Do, P. Mueller, and M. Vannucci (eds), *Bayesian Inference for Gene Expression and Proteomics*. Cambridge: Cambridge University Press.

Bansal, M. (2007) How to infer gene networks from protein profiles. *Molecular System Biology*, 3, 1–10.

Bar-Joseph, Z. (2004) Analyzing time series gene expression data. *Bioinformatics*, 20, 2493–2503.

Bar-Joseph, Z., Demaine, E., Gifford, D., Hamel, A., Srebro, N., and Jaakkola, T. (2003) k-ary clustering with optimal leaf ordering for gene expression data. *Bioinformatics*, 19, 1070–1078.

Bar-Joseph, Z., Farkash, S., Gifford, D., Simon, I., and Rosenfeld, R. (2004) Deconvolving cell cycle expression data with complementary information. *Bioinformatics* (Proceedings of ISMB) 20(Suppl. 1), I23–I30.

Barry, W.T., Nobel, A.B., and Wright, F.A (2005) Significance analysis of functional categories in geneexpression studies: a structured permutation approach. *Bioinformatics*, 21, 1943–1949.

Beal, M.J., Falciani, F., Gharamani, Z., Rangel, C., and Wild, D.L. (2005). A Bayesian approach to reconstructing genetic regulatory networks with hidden factors. *Bioinformatics*, 21, 349–356.

Beaumont, M. and Rannala, B. (2004) The Bayesian revolution in genetics. *Nat. Rev. Genet.*, 2004, 251–261.

Benjamini, Y., and Hochberg, Y. (1995) Controlling the false discovery rate: a practical and powerful approach to multiple testing. *Journal of the Royal Statistical Society Series B*, 57(1), 289–300.

Berger, J. (1985) *Statistical Decision Theory and Bayesian Analysis*. New York: Springer-Verlag.

Bernardo, J.M., and Smith, A.F.M. (1994) *Bayesian Theory*. New York: John Wiley & Sons, Ltd.

Berry, D.A., and Hochberg, Y. (2001) Bayesian perspectives on multiple comparisons. *Journal of Statistical Planning and Inference*, 82, 215–227, DOI: 10.1016/S0378-3758(99)00044-0.

Bhardwaj, N. and Lu, H. (2005) Correlation between gene expression profiles and protein–protein interactions within and across genomes. *Bioinformatics*, 21, 2730–2738.

Bhattacharjee, A., Richards, W.G., Staunton, J., Li, C., Monti, S., Vasa, P., Ladd, C., Beheshti, J., Bueno, R., Gillette, M., Loda, M., Weber, G., Mark, E.J., Lander, E.S., Wong, W., Johnson, B.E., Golub, T.R., Sugarbaker, D.J., and Meyerson, M. (2001) Classification of human lung carcinomas by mRNA expression profiling reveals distinct adenocarcinoma subclasses. *Proc. Natl. Acad. Sci. USA*, 98(24), 13790–13795.

Bickel, D.R. (2004) Error-rate and decision-theoretic methods of multiple testing: Which genes have high objective probabilities of differential expression? *Statistical Applications in Genetics and Molecular Biology*, 3(1), 8.

Binder, D.A. (1978) Bayesian cluster analysis. *Biometrika*, 65(1), 31–38.

Binder, D.A. (1981) Approximations to Bayesian clustering rules I. *Biometrika*, 68(1), 275–285.

Blackwell, D. (1973) Discreteness of Ferguson selection. *Annals of Statistics*, 1, 356–358.

Blackwell, D., and MacQueen, J.B. (1973). Ferguson distributions via Polya urn schemes. *Annals of Statistics*, 1, 353–355.

Boettcher, S. and Dethlefsen, C. (2003) Deal: A package for learning Bayesian networks. *Journal of Statistical Software*, 8(20).

Bolstad, B.M. (2006) *Pre-processing Microarray Data in Fundamentals of Data Mining for Genomics and Proteomics*. Dubitzky W, Granzow M, Berrar DP (Eds.), Springer, 2006

Bolstad, B.M., Irizarry, R.A., Astrand, M., and Speed, T.P. (2003) A comparison of normalization methods for high density oligonucleotide array data based on bias and variance. *Bioinformatics*, 19(2), 185–193.

Box, G.E.P. (1980) Sampling inference, Bayes inference, and robustness in the advancement of learning. In J.M. Bernardo, M.H. DeGroot, D.V. Lindley, and A.F.M. Smith (eds), *Bayesian Statistics*, pp. 366–381. Oxford: Oxford University Press.

Brazma, A., Hingamp, P., Quakenbush, J. et al. (2001) Minimum Information about Micorarray Experiment (MAIME) – towards standards for microarray data. Nature Genetics, 29, 365–371.

Breiman L., Friedman, J.H., Olshen, R., and Stone, C.J. (1984). *Classification and Regression Trees*. Belmont, CA: Wadsworth.

Brettschneider, J., Collin, F., Bolstad, B.M., and Speed, T.P. (2008) Quality assessment for short oligonucleotide microarray data. *Technometrics* 50(3), 241–264.

Broet, P., Richardson, S., and Radvanyi, F. (2002) Bayesian hierarchical model for identifying changes in gene expression from microarray experiments. *Journal of Computational Biology*, 9, 671–683.

Bueno-de-Mesquita, J.M., van Harten, W.H., Retel, V.P., Van't Veer, L.J., van Dam, F.S., Karsenberg, K., Douma, K.F., van Tinteren, H., Peterse, J.L., Wesseling, J., Wu, T.S., Atsma, D., Rutgers, E.J., Brink, G., Floore, A.N., Glas, A.M., Roumen, R.M., Bellot, F.E., van Krimpen, C., Rodenhuis, S., van de Vijver, M.J., and Linn, S.C. (2007) Use of 70-gene signature to predict prognosis of patients with node-negative breast cancer: a prospective community-based feasibility study (RASTER). *Lancet Oncol.*, 8(12), 1079–1087.

Bush, C. and MacEachern, S. (1996) A semiparametric Bayesian model for randomised block designs. *Biometrika*, 83, 275–285.

Cardoso, F., Van't Veer, L., Rutgers, E., Loi, S., Mook, S., and Piccart-Gebhart, M.J. (2008) Clinical application of the 70-gene profile: the MINDACT trial. *J Clin Oncol.*, 26(5), 729–735.

Carlin, B.P., and Louis, T.A. (2000) *Bayes and Empirical Bayes Methods for Data Analysis*, 2nd edition Boca Raton, FL: Chapman & Hall/CRC Press.

Chen, Y., Dougherty, E.R., and Bittner, M.L. (1997). Ratio-based decisions and the quantitative analysis of cDNA microarray images. *Journ. Biomedical Optics*, 4, 364–374.

Chen, J., and Sarkar, S.K. (2005) A bayesian determination of threshold for identifying differentially expressed genes in microarray experiments. *Statistics in Medicine*, 25, 3174–3189.

Chib, S., and Greenberg, E. (1998). Analysis of multivariate probit models. *Biometrika*, 85, 347–361.

Chi, Y., Ibrahim, J., Bissahoyo, A., and Threadgill, D. (2007) Bayesian hierarchical modeling for time course microarray experiments. *Biometrics*, 63, 496–504.

Chi, Z. (2007) Sample size and positive false discovery rate control for multiple testing, *Elec. Journ. of Stat.*, 1, 77–118.

Chipman, H., George, E.I., and McCulloch, R.E. (2001) The practical implementation of Bayesian model selection (Pkg: p65-134) *Model selection [Institute of Mathematical Statistics lecture notes-monograph series 38*, 65–116. Fountain Hills, AZ: IMS Press.

Churchill GA (2002) Fundamentals of experimental design for cDNA microarrays. *Nature Genetics*, 32, 490–495.

Clyde, M., DeSimone, H., and Parmigiani, G. (1996) Prediction via orthogonalized model mixing. *Journal of the American Statistical Association*, 91, 1197–1208.

Cooper, G., and Herskovitz, E. (1992) A Bayesian method for the induction of probabilistic networks from data. *Machine Learning*, 9, 309–347.

Cowles, M.K., and Carlin, B.P. (1996) Markov chain Monte Carlo convergence diagnostics: a comparative review. *Journal of the American Statistical Association*, 91, 883–904.

Cox, D.R. (1972) Regression models and life tables. *Journal of the Royal Statistical Society Series B*, 34, 187–220.

Craig, P., Goldstein, M., Seheult, A., and Smith, J. (1998) Constructing partial prior specifications for models of complex physical systems. *The Statistician*, 47, 37–53.

Cristianini, N., and Shawe-Taylor, J. (2000) *An Introduction to Support Vector Machines*. Cambridge: Cambridge University Press.

Curtis, R.K., Oresic, M., and Vidal-Puig, A. (2005) Pathways to the analysis of microarray data. *Trends in Biotechnology*, 23, 429–435.

Dahl, D.B., and Newton, M.A. (2007). Multiple hypothesis testing by clustering treatment effects. *Journal of the American Statistical Association*, 102, 517–526.

Dawid, A.P., and Lauritzen, S. (1995) Hyper Markov laws in the statistical analysis of decomposable graphical models. *Annals of Statistics*, 21, 1272–1317.

de Finetti, B. (1930) Funzione caratteristica di un fenomeno aleatoria. *Men. Acad. Naz. Lincei.*, 4, 86–133.

de Finetti, B. (1937) La prévision: ses lois logiques, ses source subjectives. Annales de l'Institut Henri Poincaré, 7(1), 1–68.   Reprinted (1964) in H.E. Kyburg and H.E. Smokler (eds), *Studies in Subjective Probability*. New York: John Wiley & Sons, Inc.

de Finetti, B. (1963) La décision et les probabilités. *Rev. Roumine Math. Pures Appl.*, 7, 405–413.

de Finetti, B. (1964) Probabilità. Reprinted (1972) as: Conditional probabilities and decision theory. In *Probability, Induction and Statistics: The Art of Guessing*. New York: John Wiley & Sons, Inc.

DeGroot M.H. (1970) *Optimal statistical decisions*. New York: McGraw-Hill.

Denison, D.G.T., Holmes, C.C., Mallick, B.K., and Smith, A.F.M. (2002) *Bayesian Methods for Nonlinear Classification and Regression*. Chichester: John Wiley & Sons, Ltd.

DeRisi, J., Penland, L., Brown, P.O., Bittner, M.L., Meltzer, P.S., Ray, M., Chen, Y., Su, Y.A., and Trent, J.M. (1996) Use of a cDNA microarray to analyse gene expression patterns in human cancer. *Nature Genetics*, 14(4), 457–460.

DeRisi, J. L., Iyer, V. R., and Brown, P. O. (1997) Exploring the metabolic and genetic control of gene expression on a genomic scale. *Science*, 278, 680–685.

Desmedt, C., Haibe-Kains, B., Wirapati, P., Buyse, M., Larsimont, D., Bontempi, G., Delorenzi, M., Piccart, M., and Sotiriou, C. (2008) Biological processes associated with breast cancer clinical outcome depend on the molecular subtypes. *Clin Cancer Res.*, 14(16), 5158–5165.

Devroye L. (1986) *Non-Uniform Random Variate Generation*. New York: Springer-Verlag.

Dey, D., Muller, P., and Sinha, D. (1998) *Practical Nonparametric and Semiparametric Bayesian Statistics*. Berlin: Springer-Verlag.

Dey D. K. Ghosh S. and Mallick B. K. (2000). *Generalized Linear Models: a Bayesian Perspective*. New York: Marcel Dekker.

Dhesi, G.S., and Jones, R.C. (1990) Asymptotic corrections to the Wigner semicircular eigenvalue spectrum of a large real symmetric random matrix using the replica method. *J. Phys. A: Math. Gen*, 23, 5577–5599. Printed in the UK.

Do, K.A., Mueller, P., and Tang, F. (2005) A nonparametric Bayesian mixture model for gene expression. *Applied Statistics*, 54(3), 1–18.

Do K.A., Mueller P., and Vannucci M. (eds) (2006) *Bayesian Inference for Gene Expression and Proteomics*. Cambridge: Cambridge University Press.

Dobbin, K.K., Beer, D.G., Meyerson, M., Yeatman, T.J., Gerald, W.L., Jacobson, J.W., Conley, B., Buetow, K.H., Heiskanen, M., Simon, R.M., Minna, J.D., Girard, L., Misek, D.E., Taylor, J.M., Hanash, S., Naoki, K., Hayes, D.N., Ladd-Acosta, C., Enkemann, S.A., Viale, A., and Giordano, T.J. (2005) Interlaboratory comparability study of cancer gene expression analysis using oligonucleotide microarrays. *Clin Cancer Res.*, 11, 565–572.

Dobra, A., Jones, B., Hans, C., Nevis, J. and west, M. (2004) Sparse graphical models for exploring gene expression data. *Journal of Multivariate Analysis*, 90, 196–212.

Dougherty ER (2001) Small sample issues for microarray-based classification. *Comp Funct Genomics.*, 2(1): 28–34.

Dudoit, S. and Fridlyand, J. Classification in microarray experiments. (Analysis of microarray experiments, Chapman & Hall/CRC, 2003, edited by T. P. Speed).

Dudoit S., Y. H. Yang, T. P. Speed, and M. J. Callow (2002). Statistical methods for identifying differentially expressed genes in replicated cDNA microarray experiments. *Statistica Sinica*, Vol. 12, No. 1, p. 111–139.

Dudoit S., Shaffer, J.P. and Boldrick, J.C. (2003). Multiple hypothesis testing in microarray experiments. *STATSCI*, Vol. 18, No. 1, p. 71–103.

Dudoit S., van der Laan M.J., and Pollard, K.S. (2004). Multiple testing. Part I. Single-step procedures for control of general Type I error rates. *Statistical Applications in Genetics and Molecular Biology*, Vol. 3, No. 1, Article 13.

Duggan, D.J., Bittner, M.L., Chen, Y., Meltzer, P.S., and Trent, J.M. (1999). Expression profiling using cDNA microarrays. *Nature Genetics*, 21, 10–14.

Duncan, D.B. (1965) A bayesian approach to multiple comparisons. *Technometrics*, 7, 171–222.

Efron, B. and Tibshirani, R. (2002) Empirical Bayes methods and false discovery rates for microarrays. *Genetic Epidemiology*, 23, 70–86.

Efron, B. and Tibshirani, R. (2006) On testing the significance of sets of genes. Technical Report, Standford University.

Efron, B., Tibshirani, R., Storey, J.D., and Tusher, V. (2001) Empirical Bayes analysis of a microarray experiment. *Journal of the American Statistical Association*, 96, 1151–1160.

Eisen, M.B., Spellman, P.T., Brown, P.O., and Botstein, D. (1998) Cluster analysis and display of genome-wide expression patterns. *Proc. Natl. Acad. Sci. USA*, 95(25), 14863–14868.

Escobar, M.D. (1994) Estimating normal means with a Dirichlet process prior. *Journal of the American Statistical Association*, 89, 268–277.

Escobar, M.D. and West, M. (1995) Bayesian density estimation and inference using mixtures. *Journal of the American Statistical Association*, 90, 577–588.

Everitt, B. (1993) *Cluster Analysis*. London: Edward Arnold.

Fan, C., Oh, D.S., Wessels, L., Weigelt, B., Nuyten, D.S., Nobel, A.B., Van't Veer, L.J., and Perou, C.M. (2006) Concordance among gene-expression-based predictors for breast cancer. *New England Journal of Medicine*, 355(6), 560–569.

Fan, J., Tam, P., Woude, G.V., and Ren, Y. (2004) Normalization and analysis of cDNA microarrays using within-array replications applied to neuroblastoma cell response to a cytokine. *Proc. Natl. Acad. Sci. USA*, 101(5), 1135–1140.

Fare, T.L., Coffey, E.M., Dai, H., He, Y.D., Kessler, D.A., Kilian, K.A., Koch, J.E., LeProust, E., Marton, M.J., Meyer, M.R., Stoughton, R.B., Tokiwa, G.Y., and Wang, Y. (2003) Effects of atmospheric ozone on microarray data quality. *Anal. Chem.*, 75, 4672–4675.

Farmer, P., Bonnefoi, H., Becette, V., Tubiana-Hulin, M., Fumoleau, P., Larsimont, D., MacGrogan, G., Bergh, J., Cameron, D., Goldstein, D., Duss, S., Nicoulaz, A.-L., Brisken, C., Fiche, M., Delorenzi, M., and Iggo R. (2005) Identification of molecular apocrine breast tumours by microarray analysis. *Oncogene*, 24, 4660–4671.

Ferguson, T.S. (1973) A Bayesian analysis of some nonparametric problems. *Annals of Statistics*, 1, 209–230.

Fisher, R.A. (1922) On the interpretation of ξ^2 from contingency tables, and the calculation of P. *Journal of the Royal Statistical Society*, 85(1), 87–94.

Fix, E., and Hodges, J.L. (1951) Discriminatory analysis-nonparametric discrimination: consistency properties. US Air Force School of Aviation Medicine, Randolph Field, TX.

Fraley, C., and Raftery, A.E. (2002) Model-based clustering, discriminant analysis, and density estimation. *Journal of the American Statistical Association*, 97(458), 611–631.

Freedman, D.A. (1963) On the asymptotic behavior of Bayes estimate in the discrete case. *Annals of Mathematical Statistics*, 34, 1386–1403.

Friedman, J.H. (1991). Multivariate adaptive regression splines (with discussion). *Annals of Mathematical Statistics*, 19, 1–141.

Friedman, J. H. (1997) On bias, variance, 0/1 – loss, and the curse-of-dimensionality. *Data Mining and Knowledge Discovery*, 1(1), 55–77.

Friedman, N. (2004) Inferring cellular networks using probabilistic graphical models. *Science*, 30, 799–805.

Friedman, N., and Koller, D. (2003) Being Bayesian about network structure: a Bayesian approach to structure discovery in ayesian networks. *Machine Learning*, 50, 95–125.

Frühwirth-Schnatter, S. (2001) MCMC estimation of classical and dynamic switching and mixture models. *Journal of the American Statistical Association*, 96, 194–209.

Fu, J., Jeffrey, S.S. (2007) Transcriptomic signatures in breast cancer. *Mol. Biosyst.*, 3(7), 466–472.

Garthwaite, P.H., and Dickey, J.M. (1996) Quantifying and using expert opinion for variable-selection problems in regression. *Chemometrics and Intelligent Laboratory Systems*, 35, 1–26; discussion 27–43.

Gasch, A., Spellman, P., Kao, C., Carmel-Harel, O., Eisen, M., Storz, G., Botstein, D. and Brown, P. (2000) Genomic expression programs in the response of yeast cells to environmental changes. *Mol. Biol. Cell*, 11, 4241–4257.

Geiger, D. and Heckerman, D. (1997) A characterization of Dirichlet distributions through local and global independence. *Annals of Statistics*, 25, 1344–1368.

Gelfand A. (2000) Gibbs sampling. *Journal of the American Statistical Association*, 95, 1300–1304.

Gelfand, A. (1996). Model determination using sampling-based methods. In W.R. Gilks, S. Richardson, and D.J. Spiegelhalter (eds), *Markov Chain Monte Carlo in Practice*. London: Chapman & Hall.

Gelfand A.E. and Smith A.F.M. (1990) Sampling-based approaches to calculating marginal densities. *Journal of the American Statistical Association*, 85(410), 398–409.

Gelfand, A., Hills, S.E., Racine-Poon, A., and Smith, A.F.M. (1990) Illustration of Bayesian inference in normal data models using Gibbs sampling. *Journal of the American Statistical Association*, 85, 972–985.

Geller, S.C., Gregg, J.P., Hagerman, P., and Rocke, D.M. (2003) Transformation and normalization of oligonucleotide microarra data. *Bioinformatics*, 19(14), 1817–1823.

Gelman, A., and Rubin, D.B. (1992) Inference from iterative simulation using multiple sequences. *Stat. Sci.*, 7, 457–511.

Geman, S., and Geman, D. (1984) Stochastic relaxation, Gibbs distributions, and the Bayesian restoration of images. *IEEE Transactions on Pattern Analysis and Machine Intelligence*, 6, 721–741.

Genovese, C., and Wasserman, L. (2002) Operating characteristics and extensions of the false discovery rate procedure. *Journal of the Royal Statistical Society, Series B*, 64(3), 499–517.

Genovese, C., and Wasserman, L. (2004) A stochastic process approach to false discovery control. *Annals of Statistics*, 32, 1035–1061.

Gentleman, R.C., Carey, V.J., Bates, D.M., Bolstad, B., Dettling, M., Dudoit, S., Ellis, B., Gautier, L., Ge, Y., Gentry, J., Hornik, K., Hothorn, T., Huber, W., Iacus, S., Irizarry, R., Leisch, F., Li, C., Maechler, M., Rossini, A.J., Sawitzki, G., Smith, C., Smyth, G., Tierney, L., Yang, J.Y., and Zhang, J. (2004) Bioconductor: open software development for computational biology and bioinformatics. *Genome Biol.*, 5(10), R80.

Geoman, J.J., van de Geer, S., de Kort, F., and van Houwelingen, H.C. (2004) A global test for groups of genes: testing association with a clinical outcome. *Bioinformatics*, 20, 93–99.

George, E.I., and McCulloch, R.E. (1993) Variable selection via Gibbs sampling. *Journal of the American Statistical Association* 88, 881–889.

George, E.I., and McCulloch, R.E. (1997) Approaches for Bayesian variable selection. *Statistica Sinica*, 7, 339.

Geyer, C.J. (1992) Practical Markov chain Monte Carlo. *Statistical Science*, 7(4), 473–483.

Gilks, W.R., and Wild, P. (1992) Adaptive rejection sampling for Gibbs sampling. *Applied Statistics*, 41, 337–348.

Gold, D.L., Wang, J., and Coombes, K.R. (2005) Inter-gene correlation on oligonucleotide arrays: how much does normalization matter? *American Journal of Pharmacogenomics*, 5, 271–279.

Gold, D.L., Coombes, K.R., Wang, J. and Mallick, B. (2007) Enrichment analysis in high-throughput genomics – accounting for dependency in the NULL. *Briefings in Bioinformatics*, 8, 71–77.

Goldstein, H., Rasbash, J., Yang, M., Woodhouse, G., Pan, H., Nuttall, D., and Thomas, S. (1993) A multilevel analysis of school examination results. *Oxford Review of Education*, 19(4), 425–433.

Golub, T.R., Slonim, D.K., Tamayo, P., Huard, C., Gaasenbeek, M., Mesirov, J.P., Coller, H., Loh, M.L., Downing, J.R., Caligiuri, Bloomfoiel, C.D., and Lander E.S. (1999) Molecular classification of cancer: class discovery and class prediction by gene expression monitoring. *Science*, 286, 531–537.

Good, I.J. (1988) The interface between statistics and philosophy of science (with discussion). *Statistical Science*, 3, 386–398.

Gottardo, R., Raftery, A.E., Yee Y.K., and Bumgarner, R.E. (2006) Bayesian robust inference for differential gene expression in microarrays with multiple samples. *Biometrics*, 62, 10–18.

Green, P.J. (1995). Reversible jump Markov chain Monte Carlo computation and Bayesian model determination. *Biometrika*, 82, 711–732.

Guo, X., Qi, H., Verfaillie, C., and Pan, W. (2003). Statistical significance analysis of longitudinal gene expression data. *Bioinformatics*, 19, 1628–1635.

Haab, B.B., Dunham, M.J., and Brown, P.O. (2001) Protein microarrays for highly parallel detection and quantitation of specific proteins and antibodies in complex solutions. *Genome Biology*, 2(2), research 0004.1–0004.13.

Hastie, T., Tibshirani, R., Botstein, D., and Brown, P. (2001) Supervised harvesting of expression trees. *Genome Biology*, 2, research 0003.1–0003.12.

Hastings W. (1970) Monte Carlo sampling methods using Markov chains and their applications. *Biometrika*, 57, 97–110.

Heard, N., Holmes, C., Stephens, D., Hand, D. and Dimopoulos, G. (2005) A Bayesian coclustering of Anopheles gene expression time series response to multiple immune challenges. *Proc. Natl. Acad. Sci. USA*, 102, 16939–16944.

Heckerman, D., Geiger, D., and Chickering, D. (1995) Learning Bayesian networks: the combination of knowledge and statistical data. *Machine Learning*, 20, 197–243.

Hedenfalk, I., Duggan, D., Chen, Y., Radmacher, M., Bittner, M., Simon, R., Meltzer, P., Gusterson, B., Esteller, M., Kallioniemi, O.P., Wilfond, B., Borg, A., and Trent J. (2001). Gene expression profiles in hereditary breast cancer. *New England Journal of Medicine*, 344, 539–548.

Herbrich, R. (2002) *Learning Kernel Classifiers*. Cambridge, MA: MIT Press.

Hess, K.R., Zhang, W., Baggerly, K.A., Stivers, D.N., and Coombes, K.R. (2001) Microarrays: handling the deluge of data and extracting reliable information. *Trends Biotechnol.*, 19(11), 463–468.

Hodges, J.S. (1998) Some algebra and geometry for hierarchical models, applied to diagnostics. *Journal of the Royal Statistical Society, Series B*, 60, 497–536.

Holmes, C.C., and Denison, D.G.T. (2003). Classification with Bayesian MARS. *Machine Learning*, 50, 159–173.

Holmes, C.C. and Held, K. (2006) Bayesian auxiliary variable models for binary and polychotomous regression. *Bayesian Analysis*, 1(1), 145–168.

Holter, N.S., Maritan, A., Cieplak, M., Fedoroff, N., and Banavar, J. (2001) Dynamic modeling of gene expression data. *Proc. Natl. Acad. Sci. USA*, 98, 1693–1698.

Huang, S. (1999). Gene expression profiling, genetic networks and cellular states; an integrating concept for tumorigenesis and drug discovery. *J. Mol. Med.*, 77, 469–480.

Huang, J., Wang, D., and Zhang, C. (2005) A two-way semilinear model for normalization and analysis of cDNA microarray data. *Journal of the American Statistical Association*, 471, 814–829.

Hubert, L., and Arabie, P. (1985) Comparing partitions. *Journal of Classification*, 2, 193–218.

Husmeier, D. (2003) Sensitivity and specificity of inferring genetic regulatory interactions from microarray experiments with dynamic bayesian networks. *Bioinformatics*, 19, 2271–2282.

Ibrahim, J.G., Chen, M.-H., and Gray, R.J. (2002) Bayesian models for gene expression with DNA microarray data, *Journal of the American Statistical Association*, 97, 88–99.

Ideker, T., Thorsson, V., Siegel, A.F., and Hood, L.R. (2000). Testing for differentially-expressed genes by maximum-likelihood analysis of microarray data. *Journal of Computational Biology*, 6, 805–817.

Ideker, T., Thorsson, V., Ranish, J.A., Christmas, R., Buhler, J., Eng, J.K., Bumgarner, R., Goodlett, D.R., Aebersold, R., and Hood, L. (2001) Integrated genomic and proteomic analyses of a systematically perturbed metabolic network. *Science*, 292(5518), 929–934.

Irizarry, R.A., Bolstad, B.M., Collin, F., Cope, L.M., Hobbs, B., and Speed, T.P. (2003) Summaries of Affymetrix GeneChip probe level data. *Nucleic Acids Research*, 31(4).

Irizarry, R.A., Warren, D., Spencer, F., Kim, I.F., Biswal, S., Frank, B.C., Gabrielson, E., Garcia, J.G., Geoghegan, J., Germino, G., Griffin, C., Hilmer, S.C., Hoffman, E., Jedlicka, A.E., Kawasaki, E., Martrnez-Murillo, F., Morsberger, L., Lee, H., Petersen,

D., Quackenbush, J., Scott, A., Wilson, M., Yang, Y., Ye, S.Q., and Yu, W. (2005) Multiple-laboratory comparison of microarray platforms. *Nature Methods*, 2, 345–349.

Ishwaran, H., and Rao, J.S. (2003). Detecting differentially expressed genes in microarrays using Bayesian model selection. *Journal of the American Statistical Association*, 98, 438–455.

Ishwaran, H., and Rao, J.S. (2005). Spike and slab gene selection for multigroup microarray data. *Journal of the American Statistical Association*, 100, 764–780.

Ishwaran H., Rao, J.S., and Kogalur, U.B. (2006). BAMarray™: Java software for Bayesian analysis of variance for microarray data. *BMC Bioinformatics*, 7: 59.

Jasra, A., Holmes, C.C., and Stephens, D.A. (2005) Markov chain Monte Carlo methods and the label switching problem in Bayesian mixture modeling. *Statist. Sci.*, 20(1), 50–67.

Jeffreys, H. (1946) An invariant form for the prior probability in estimation problems. *Proceedings of the Royal Society of London, Series A: Mathematical, Physical and Engineering Sciences*, 186, 453–461.

Jensen, F.V. (1996) *An Introduction to Bayesian Networks*. London: UCL Press.

Ji, Y., Tsui, K.-W., and Kim, K.M. (2006) A two-stage empirical Bayes method for identifying differentially expressed genes. *Computational Statistics & Data Analysis*, 50, 3592–3604.

Jiang, Z., and Gentleman, R. (2007) Extensions to gene set enrichment. *Bioinformatics*, 23(3), 306–313.

Kadane, J. and Wolfson, L. (1998) Experience in elicitation. *The Statistician*, 47, 3–19.

Kalbfleisch, J. (1978) Non-parametric Bayesian analysis of survival time data. *Journal of the Royal Statistical Society, Series B*, 40, 214–221.

Kalbfleisch, J., and Prentice, R. (1980) *The Statistical Analysis of Failure Time Data*. New York: John Wiley & Sons, Inc.

Kass, R., Tierney, L., and Kadane J. (1988) Asymptotics in Bayesian computation. In J.M. Bernardo, M.H. DeGroot, D.V. Lindley, and A.F.M. Smith (eds), *Bayesian Statistics 3*, pp. 263–278. Oxford: Oxford University Press.

Kass, R., Tierney, L., and Kadane J. (1989). Fully exponential Laplace approximations to expectations and variances of nonpositive functions. *Journal of the American Statistical Association*, 84, 710–716.

Kerr, M.K., and Churchill, G.A. (2001a) Experimental design for gene expression microarrays. *Biostatistics*, 2(2), 183–201.

Kerr, M.K., and Chruchill, G.A. (2001b) Statistical design and the analysis of gene expression microarray data. *Genetics Research*, 77, 123–128.

Kerr, M.K., Martin, M., and Churchill, G.A. (2000) Analysis of variance for gene expression microarray data. *Journal of Computational Biology*, 7(6), 819–837.

Kim, S., Imoto, S., and Miyano, S. (2003) Inferring gene networks from time series microarray data using dynamic Bayesian networks. *Briefings in Bioinformatics*, 4, 228–235.

Kim, S.-Y., and Volsky, D.J. (2005) PAGE: Parametric Analysis of Gene Set Enrichment. *BMC Bioinformatics*, 6, 144.

Klebanov, L., Glazko, G., Salzman, P., Yakovlev, A., and Xiao, Y. (2007) A multivariate extension of the gene set enrichment analysis. *J. Bioinform. Comput. Biol.*, 5(5), 1139–1153.

Klevecz, R.R., Bolen, J., Forrest, G., and Murray D.B. (2004). A genomewide oscillation in transcription gates DNA replication and cell cycle. *Proc. Natl. Acad. Sci. USA*, 101, 1200–1205.

Koscielny S (2008) Critical review of microarray-based prognostic tests and trials in breast cancer. *Curr. Opin. Obstet. Gynecol.*, 20(1), 47–50.

Kooperberg, C., Bose, S., and Stone, C. J. (1997). Polychotomous regression. *Journal of the American Statistical Association*, 93, 117–127.

Lander, E.S., and Weinberg, R.A. (2000). Journey to the center of biology. *Science*, 287, 1777–1782.

Larranaga, P., Kuijpers, C., Murga, R., and Yurramendi, Y. (1996) Learning Bayesian network structures by searching for the best ordering with genetic algorithms. *IEEE Transactions on Pattern Analysis and Machine Intelligence*, 26, 487–493.

Lauritzen, S. L. (1996). *Graphical Models*. Oxford: Oxford University Press.

Lee, K., and Mallick, B. (2004) Bayesian methods for variable selection in survival models with application to DNA microarray data. *Sankhya*, 66, 756–778.

Lee, K.Y., Sha N., Doughetry, E.R., Vannucci, M., and Mallick, B.K. (2003). Gene selection: a Bayesian variable selection approach. *Bioinformatics*, 19, 90–97.

Lee, M.-L.T. (2004) *Analysis of Microarray Gene Expression Data*. Boston: Kluwer Academic.

Lewin, A., Bochkina, N., and Richardson, S. (2007). Fully Bayesian mixture model for differential gene expression: Simulations and model checks. *Statistical Applications in Genetics and Molecular Biology*, 6.

Li, C., and Wong, W.H. (2001) Model-based analysis of oligonucleotide arrays: model validation, design issues and standard error application. *Genome Biology*. 2(8), research 0032.I–0032.II

Li, H., and Gui, J. (2004) Partial Cox regression analysis for high-dimensional microarray gene expression data. *Bioinformatics*, 20, 208–215.

Li, Y., Campbell, C., and Tipping, M. (2002). Bayesian automatic relevance determination algorithms for classifying gene expression data. *Bioinformatics*, 18, 1332–1339.

Liao, J.C., Boscolo, R., Yang, Y.-L., Tran, L.M., Sabatti, C., and Roychowdhury V.P. (2003) Network component analysis: Reconstruction of regulatory signals in biological systems. *Proc. Natl. Acad. Sci. USA*, 100, 15522–15527.

Lindley, D.V. (1957) A statistical paradox. *Biometrika*, 45, 533–534.

Lindley, D.V., and Smith, A.F.M. (1972) Bayes estimates for the linear model (with discussion). *Journal of the Royal Statistical Society, Series B*, 34, 1–41.

Lu, Y., Liu, P.Y., Xiao, P., and Deng, H.W. (2005) Hotelling's T^2 multivariate profiling for detecting differential expression in microarrays. *Bioinformatics*, 21(14), 3105–3113.

Luan, Y., and Li, H. (2004). Model-based methods for identifying periodically expressed genes based on time course microarraygene expression data. *Bioinformatics*, 20, 332–339.

MacEachern, S. (1994) Estimating normal means with a conjugate style Dirichlet process prior. *Communications in Statistics, B*, 23, 727–741.

Mallick, B.K., Ghosh, D., and Ghosh, M. (2005) Bayesian classification of tumors using gene expression data. *Journal of the Royal Statistical Society, Series B*, 67, 219–234.

Ma, S., Song, X., and Huang, J. (2007) Supervised group lasso with applications to microarray data analysis. *BMC Bioinformatics*, 8, 60.

Madigan, D., and Raftery, A. (1994) Model selection and accounting for model uncertainty in graphical models using Occam's window. *Journal of the American Statistical Association*, 89, 1535–1546.

Mardia, K.V., Kent, J.T., and Bibby, J.M. (1979) *Multivariate Analysis*. London: Academic Press.

MAQC Consortium (2006) The MicroArray Quality Control (MAQC) project shows inter- and intraplatform reproducibility of gene expression measurements. *Nature Biotech.*, 24, 1151–1161.

McCall, M.N., and Irizarry, R.A. (2008) Consolidated strategy for the analysis of microarray spike-in data. *Nucleic Acids Research*, 36(17), e108.

McCullagh, P., and Nelder, J.A. (1989) *Generalized Linear Models*, 2nd edition. London: Chapman & Hall.

Medvedovic, M., and Sivaganesan, S. (2002) Bayesian infinite mixture model-based clustering of gene expression profiles. *Bioinformatics*, 18, 1194–1206.

Mestl, T., Plahte, E., and Omholt, S. (1995). A mathematical framework for describing and analyzing gene regulatory network. *Journal of Theoretical Biology*, 176, 291–300.

Metropolis, N., Rosenbluth, A., Rosenbluth, M., Teller, A.H., and Teller, E. (1953) Equations of state calculations by fast computing machines. *Journal of Chemical Physics*, 21, 1087–1091.

Millikan, R.C., Newman, B., Tse, C.K., Moorman, P.G., Conway, K., Dressler, L.G., Smith, L.V., Labbok, M.H., Geradts, J., Bensen, J.T., Jackson, S., Nyante, S., Livasy, C., Carey, L., Earp, H.S., Perou, C.M. (2007) Epidemiology of basal-like breast cancer. *Breast Cancer Res Treat.*, 109(1), 123–139.

Mitchell, T.J., and Beauchamp, J.J. (1988). Bayesian variable selection in linear regression. *Journal of the American Statistical Association*, 83, 1023–1036.

Moloshok, T.D., Klevecz, R.R., Grant, J.D., Manion, F.J., Speier IV, W.F., and Ochs, M.F. (2002) Application of Bayesian decomposition for analyzing microarray data. *Bioinformatics*, 18, 566–575.

Mootha, V.K., Lindgren, C.M., Eriksson, K.F., Subramanian, A., Sihag, S., Lehar, J., Puigserver, P., Carlsson, E., Ridderstrale, M., Laurila, E., Houstis, N., Daly, M.J., Patterson, N., Mesirov, J.P., Golub, T.R., Tamayo, P., Spiegelman, B., Lander, E.S., Hirschhorn, J.N., Altshuler, D., and Groop, L.C. (2003) PGC-1alpha-responsive genes involved in oxidative phosphorylation are coordinately downregulated in human diabetes. *Nature Genetics*, 34, 267–273.

Morris, C.N. (1983) Natural exponential families with quadratic variance functions: statistical theory. *Annals of Statistics*, 11, 515–529.

Muller, P., Parmigiani, G., Robert, C., and Rousseau, J. (2004) Optimal sample size for multiple testing: the case of gene expression microarray. *Journal of the American Statistical Association*, 99, 990–1001.

Murphy, K., and Mian, S. (1999). Modelling gene expression data using dynamic Bayesian networks. Technical report. Computer Science Division, University of California, Berkeley.

Nagaraj, V.H., O'Flanagan, R.A., Bruning, A.R., Mathias, J.R., Vershon, A.K., and Sengupta, A.M. (2004) Combined analysis of expression data and transcription factor binding sites in the yeast genome. *BMC Genomics*, 5(1), 59.

Nau, G., Richmond, J., Schlesinger, A., Jennings, E., Lander, E., and Young, R. (2002) Human macrophage activation programs induced by bacterial pathogens. *Proc. Natl. Acad. Sci. USA*, 99, 1503–1508.

Neal, R. (1992) Bayesian mixture modelling. In *Proceedings of the Workshop on Maximum Entropy and Bayesian Methods of Statistical Analysis*, 11, 197–211.

Neal, R. (2000) Markov chain methods for Dirichlet process mixture models. *Journal of Computational and Graphical Statistics*, 9, 249–265.

Newton M.A., Kendziorski C.M., Richmond, C.S., Blattner, F.R., and Tsui, K.W. (2001). On differential variability of expression ratios: Improving statistical inference about gene expression changes from microarray data. *Journal of Computational Biology*, 8, 37–52.

Newton, M.A., Quintana, F.A., den Boon, J.A., Sengupta, S., and Ahlquist, P. (2007). Random-set methods identify distinct aspects of the enrichment signal in gene-set analysis. *Annals of Applied Statistics*, 1, 85–106.

Nguyen, D., and Rocke, D. (2002) Partial least squares proportional hazard regression for application to DNA microarray survival data. *Bioinformatics*, 18, 1625–1632.

Nguyen, D., Arpat, A., Wang, N., and Carroll, R. (2002) DNA microarray experiments: biological and technological aspects. *Biometrics*, 58, 701–717.

O'Hagan, A. (1998) Eliciting expert beliefs in substantial practical applications, *The Statistician*, 47, 21–35.

O'Hagan, A. and Forster, J.J. (2004) *Bayesian Inference*, 2nd edition. London: Arnold.

Olson, J. (2004) Application of microarray profiling to clinical trials in cancer. *Surgery*, 136(3), 519–523.

Ong, I., Glasner, J., and Page, D. (2002) Modelling regulatory pathways in E. coli from time series expression profiles. *Bioinformatics*, 18(Suppl. 1), 241–248.

Pan, W. (2006) Incorporating gene functional annotations in detecting differential gene expression. *Journal of the Royal Statistical Society, Series C - Applied Statistics*, 55, 301–316.

Panda, S., Antoch, M., Miller, B., Su, A., Schook, A., Straume, M., Schultz, P., Kay, S., Takahashi, J., and Hogenesch, J. (2002) Coordinated transcription of key pathways in the mouse by the circadian clock. *Cell*, 109, 307–320.

Park, T., Yi, A.G., Lee, S. Yoo, D.H., Ahn, J.I., and Lee, Y.S. (2003). Statistical tests for identifying differentially expressed genes in time-course microarray experiments. *Bioinformatics*, 19, 694–703.

Park, P.J., Pagano, M., and Bonetti, M. (2001) A nonparametric scoring algorithm for identifying informative genes from microarray data. *Pacific Symposium on Biocomputing*, 6, 52–63.

Parmigiani, G., Garrett, E.S., Anbazhagan, R., and Gabrielson, E. (2002). A statistical framework for expression-based molecular classification in cancer (with discussion). *Journal of the Royal Statistical Society, Series B*, 64, 717–736.

Parmigiani, G., Garrett, E.S., Irizarry, R.A., and Zeger S. L. (2003). *The Analysis of Gene Expression Data: Methods and Software*. New York: Springer-Verlag.

Parzen, E. (1970) Statistical inference on time series by rkhs methods. In R. Pyke (ed.), *Proc. 12th Bienn. Sem.*, pp. 1–37. Montreal: Canadian Mathematical Congress.

Pearl, J. (1988) *Probabilistic Reasoning in Intelligent Systems: Networks of Plausible Inference*. San Francisco: Morgan Kaufmann.

Perrin, B.E., Ralaivola, L., Mazurie, A., Bottani, S., Mallet, J., and d'Alché-Buc, F. (2003) Gene networks inference using dynamic Bayesian networks. *Bioinformatics*, 19, II138–II148.

Pomeroy, S.L., Tamayo, P., Gaasenbeek, M. et al. (2002). Prediction of central nervous system embryonal tumor outcome based on gene expression. *Nature*, 415(24), 436–442.

Pontil, M., Evgeniou, T., and Poggio, T. (2000) Regularization networks and support vector machines. *Adv. Comput. Math.*, 13, 1–50.

Pounds, S., and Morris, S.W. (2003) Estimating the occurrence of false positives and false negatives in microarray studies by approximating and partitioning the empirical distribution of *p*-values. *Bioinformatics*, 19, 1236–1242.

Pounds, S. and Cheng, C. (2006) Robust estimation of the false discovery rate. *Bioinformatics*, 22(16), 1979–1987.

Press, S.J. (2003) *Subjective and Objective Bayesian Statistics*. Hoboken, NJ: Wiley-Interscience.

Pusztai, L., and Hess, K.R. (2004) Clinical trial design for microarray predictive marker discovery and assessment. *Ann. Oncol.*, 15(12), 1731–1737.

Qian, J., Dolled-Filhart, M., Lin, J., Yu, H., and Gerstein, M. (2001) Beyond synexpression relationships: local clustering of time-shifted and inverted gene expression profiles identifies new, biologically relevant interactions. *Journal of Molecular Biology*, 314(5), 1053–1066.

Quintana, F.A., and Iglesias, P.L. (2003) Bayesian clustering and product partition models. *Journal of the Royal Statistical Society, Series B*, 65, 557–574.

Raftery, A.E., Madigan, D., and Hoeting, J. A. (1997) Bayesian model averaging for linear regression models. *Journal of the American Statistical Association*, 92, 179–191.

Ramaswamy, S. and Golub, T.R. (2002) DNA microarrays in clinical oncology. *Journal of Clinical Oncology*, 20(7), 1932–1941.

Ramaswamy, S., Tamayo, P., Rifkin, R., Mukherjee, S., Yeang, C.H., Angelo, M., Ladd, C., Reich, M., Latulippe, E., Mesirov, J.P., Poggio, T., Gerald, W., Loda, M., Lander, E.S., and Golub, T.R. (2001) Multiclass cancer diagnosis using tumor gene expression signatures. *Proc. Natl. Acad. Sci. USA*, 98(26), 15149–15154.

Ramaswamy, S., Ross, K.N., Lander, E.S., and Golub, T.R. (2003) A molecular signature of metastasis in primary solid tumors. *Nature Genetics*, 33(1), 49–54.

Ramoni, M., Sebastiani, P., and Kohane, I. (2002) Cluster analysis of gene expression dynamics. *Proc. Natl. Acad. Sci. USA*, 99, 9121–9126.

Ramsay, G. (1998) DNA chips: state-of-the art. *Nature Biotechnology*, 16(1), 40–44.

Rand, W.M. (1971) Objective criteria for the evaluation of clustering methods. *Journal of the American Statistical Association*, 66, 846–850.

Ray, S., and Mallick, B.K. (2006) Functional clustering by Bayesian wavelet methods. *Journal of the Royal Statistical Society, Series B*, 68, 305–332.

Reverter, A., Barris, W., McWilliam, S, Byrne, K.A., Wang, Y.H., Tan, S.H., Hudson, N., and Dalrymple, B.P. (2005) Validation of alternative methods of data normalization in gnee co-expression studies. *Bioinformatics*, 21(7), 1112–1120.

Richardson, A.M., Woodson, K., Wang, Y., Rodriguez-Canales, J., Erickson, H.S., Tangrea, M.A., Novakovic, K., Gonzalez, S., Velasco, A., Kawasaki, E.S., Emmert-Buck,

M.R., Chuaqui, R.F., and Player, A. (2007) Global expression analysis of prostate cancer-associated stroma and epithelia. *Diagn Mol Pathol.*, 16(4), 189–197.

Richardson, S., and Green, P.J. (1997). On Bayesian analysis of mixtures with an unknown number of components (with discussion). *Journal of the Royal Statistical Society, Series B*, 59, 731–792.

Rigby, R.A. (1997) Bayesian discrimination between two multivariate normal populations with equal covariance matrices. *Journal of the American Statistical Association*, 92(439), 1151–1154.

Ripley, B.D. (1987) *Stochastic Simulation*. New York: John Wiley & Sons, Inc.

Ripley, B.D. (1996). *Pattern Recognition and Neural Networks*. New York: Cambridge University Press.

Robert, C.P. (1995) Simulation of truncated normal variables. *Statistics and Computing*, 5, 121–125.

Robert, C.P. and Casella, G. (1999). *Monte Carlo Statistical Methods*. New York: Springer-Verlag.

Rosenblatt F. (1962) *Principles of Neurodynamics*. New York: Spartan.

Rosenwald, A. et al. (2002) The use of molecular profiling to predict survival after chemotherapy for diffuse large B-cell lymphoma. *New England Journal of Medicine*, 346, 1937–1946.

Savage, L.J. (1972) *The Foundations of Statistics*. New York: Dover.

Schadt, E.E., Li, Ch., Ellis, B., and Wong, W.H. (2001) Feature extraction and normalization algorithms for high-density oligonucleotide gene expression array data. *Journal of Cellular Biochemistry*, Supplement 37, 120–125.

Schaffer, J., and Strimmer, K. (2005) An emepirical Bayes approach to inferring large-scale gene association networks. *Bioinformatics*, 21, 754–764.

Scharpf, R.B., Lacobuzio-Donahue, C.A., Sneddon, J.B., and Parmigiani, G. (2006) When should one subtract background fluorescence in 2-color microarrays? *Biostatistcs*, 8(4), 697–707.

Schena, M., Shalon, D., Davis, R., and Brown, P. (1995) Quantitative monitoring of gene expression patterns with a complementary DNA microarray. *Science*, 270, 467–470.

Schena, M., Heller, R.A., Theriault, T.P., Konrad, K., Lachenmeier, E., and Davis, R.W. (1998) Microarrays: biotechnology's discovery platform for functional genomics. *Trends Biotechnol.*, 16(7), 301–306.

Schervish, M.J., and Carlin, B.P. (1992) On the convergence of successive substitution sampling. *Journal of Computational and Graphical Statistics*, 1, 111–127.

Schliep, A., Schonhuth, A., and Steinhoff, C. (2003) Using hidden Markov models to analyze gene expression time course data. *Bioinformatics*, 19, I264–I272.

Schölkopf, B., and Smola, A. (2002) *Learning with Kernels*. Cambridge, MA: MIT Press.

Schwarz, G. (1978) Estimating the dimension of a model. *Annals of Statistics*, 6, 461–464.

Sebastiani, P., Gussoni, E., Kohane, I., and Ramoni, M. (2003) Statistical challenges in functional genomics (with discussion). *Statistical Science*, 18, 33–70.

Sebastiani, P., Abad-Grau, M., and Ramoni, M. (2005) Bayesian networks. In O. Maimon and L. Rokach (eds), *The Data Mining and Knowledge Discovery Handbook*, pp. 193–230. New York: Springer-Verlag.

Segal, E., Shapira, M., Regev, A., Koller, D., and Friedman, N. (2003) Module networks: identifying regulatory modules and their condition-specific regulators from gene expression data. *Nature Genetics*, 34, 166–176.

Sethuraman, J. (1994) A constructive definition of Dirichlet priors. *Statistica Sinica*, 4, 639–650.

Sethuraman, J., and Tiwari, R. (1982) Convergence of Dirichlet measures and the interpretation of their paramegters. In S.S. Gupta and J.O. Berger (eds), *Statistical Decision theory and related topics, III*, pp. 305–315. New York: Academic Press.

Sha, N., Tadesse, M.G., and Vannucci, M. (2006). Bayesian variable selection for the analysis of microarray data with consored outcome. *Bioinformatics*, 22(18), 2262–2268.

Shalon, D., Smith, S.J., and Brown, P.O. (1996) A DNA microarray system for analyzing complex DNA samples using two-color fluorescent probe hybridization. *Genome Research*, 6(7), 639–645.

Shipp, M.A., Ross, K.N., Tamayo, P., Weng, A.P., Kutok, J.L., Aguiar, R.C., Gaasenbeek, M., Angelo, M., Reich, M., Pinkus, G.S., Ray, T.S., Koval, M.A., Last, K.W., Norton, A., Lister, T.A., Mesirov, J., Neuberg, D.S., Lander, E.S., Aster, J.C., and Golub, T.R. (2002) Diffuse large B-cell lymphoma outcome prediction by gene-expression profiling and supervised machine learning. *Nat. Med.*, 8(1), 68–74.

Shoemaker, J., Painter, I., and Weir, B. (1999) Bayesian statistics in genetics, a guide for the uninitiated. *Trends in Genetics*, 15, 354–358.

Silverman, B.W. (1986) Density Estimation for Statistics and Data Analysis (Monographs on Statistics and Applied Probability), 1st edition. Chapman & Hall/CRC.

Singh, M., and Valtorta, M. (1995) Construction of Bayesian network structures from data: A brief survey and efficient algorithm. *International Journal of Approximate Reasoning*, 12, 111–131.

Smith, M., and Kohn, R. (1996) Nonparametric regression using Bayesian variable selection. *Journal of Econometrics*, 75, 317–344.

Sollich, P. (2001) Bayesian methods for support vector machines: evidence and predictive class probabilities. *Machine Learning*, 46, 21–52.

Specht, D. F. (1990). Probabilistic neural networks. *Neur. Networks*, 3, 109–118.

Speed, T. (ed.) (2003) Statistical Analysis of Gene Expression Microarray Data. Boca Raton, FL: Chapman & Hall/CRC.

Spiegelhalter, D., and Lauritzen, S. (1990) Sequential updating of conditional probabilities on directed graphical structures. *Networks*, 20, 157–224.

Spirtes, P., Glymour, C., and Scheines, R. (1993) *Causation, Prediction and Search*. New York: Springer-Verlag.

Spurgers, K.B., Gold, D.L., Coombes, K.R., Bohnenstiehl, N.L., Mullins, B., Meyn, R.E., Logothetis, C.J., and McDonnell, T.J. (2006) Identification of cell cycle regulatory genes as principal targets of p53-mediated transcriptional repression. *J. Biol. Chem.*, 281(35), 25134–25142.

Storey, J.D. (2002) A direct approach to false discovery rates. *Journal of the Royal Statistical Society, Series B*, 64, 479–498.

Storey, J.D. (2003) The positive false discovery rate: a Bayesian interpretation and the q-value. *Annals of Statistics*, 31(6), 2013–2035.

Storey J.D., Taylor J.E., and Siegmund D. (2004) Strong control, conservative point estimation, and simultaneous conservative consistency of false discovery rates: A unified approach. *Journal of the Royal Statistical Society, Series B*, 66, 187–205.

Storey, J., Xiao, W., Leek, J.T., Dai, J.Y., Tompkins, R.G., and Davis, R.W. (2005). Significance analysis of time course microarrayexp eriments. *Proc. Natl. Acad. Sci. USA*, 102, 12837–12842.

Storch, K., Lipan, O., Leykin, I., Viswanathan, N., Davis, F., Wong, W. and Weitz, C. (2002) Extensive and divergent circadian gene expression in liver and heart. *Nature*, 418, 78–83.

Stuart, R.O., Wachsman, W., Berry, C.C., Wang-Rodriguez, J., Wasserman, L., Klacansky, I., Masys, D., Arden, K., Goodison, S., McClelland, M., Wang, Y., Sawyers, A., Kalcheva, I., Tarin, D., and Mercola, D. (2004) In silico dissection of cell-type-associated patterns of gene expression in prostate cancer. *Proc. Natl. Acad. Sci. USA*, 101(2), 615–620.

Subramanian, A., Kuehn, H., Gould, J., Tamayo, P., and Mesirov, J.P. (2007) GSEA-P: a desktop application for gene set enrichment analysis. *Bioinformatics*, 23(23), 3251–3253.

Subramanian, A., Tamayo, P., Mootha, V.K., Mukherjee, S., Ebert, B.L., Gillette, M.A., Paulovich, A., Pomeroy, S.L., Golub, T.R., Lander, E.S., and Mesirov, J.P. (2005) Gene set enrichment analysis: a knowledge-based approach for interpreting genome-wide expression profiles. *Proc. Natl. Acad. Sci. USA*, 102, 15545–15550.

Tai, Y.C., and Speed, T.P. (2006). A multivariate empirical Bayes statistic for replicated microarray time course data. *Annals of Statistics*, 34(5), 2387–2412.

Tamada, Y., Kim, S.Y., Bannai, H., Imoto, S., Tashiro, K., Kuhara, S., and Miyano, S. (2003) Estimating gene networks from gene expression data by combining Bayesian network model with promoter element detection. *Bioinformatics*, 19, 227–236.

Tanner, M. and Wong, W. (1987) The calculation of posterior distributions by data augmentation, *Journal of the American Statistical Association*, 82, 528–540.

Thompson, K.L., and Pine, P,S. (2009) Comparison of the diagnostic performance of human whole genome microarrays using mixed-tissue RNA reference samples. *Toxicology Letters*, 186(1), 58–61.

Tibshirani, R. (1996) Regression shrinkage and selection via the lasso. *Journal of the Royal Statistical Society, Series B*, 58, 267–288.

Tierny L and Kadane JB (1986) Accurate approximations for posterior moments and marginal densities. *Journal of the American Statistical Association*, 81, 82–86.

Tözeren, A. and Byers, S. (2004) *New Biology for Engineers and Computer Scientists*. Upper Saddle River, NJ: Pearson Prentice Hall.

Tusher, V.G., Tibshirani, R., and Chu, G. (2001) Significance analysis of microarrays applied to the ionizing radiation response. *Proc. Natl. Acad. Sci. USA*, 98(9), 5116–5121.

van der Laan, M., and Bryan, J. (2001) Gene expression analysis with the parametric bootstrap. *Biostatistics*, 2, 445–461.

Vapnik, V.N. (2000) *The Nature of Statistical Learning Theory*, 2nd edition. New York: Springer.

Wachi S., Yoneda K., and Wu, R. (2005) Interactome-transcriptome analysis reveals the high centrality of genes differentially expressed in lung cancer tissues. *Bioinformatics* 21: 4205–4208.

Wahba, G. (1990). *Spline Models for Observational Data*. Philadelphia: Society for Industrial and Applied Mathematics.

Wahba, G. (1999) Support vector machines, reproducing kernel Hilbert spaces and the randomized GACV. In B. Schölkopf, C. Burges and A. Smola (eds), *Advances in Kernel Methods*, pp. 69–88. Cambridge, MA: MIT Press.

Wakefield J.C., Smith, A., Racine-Poon, A., and Gelfand A. (1994) Bayesian analysis of linear and non-linear population models by using the Gibbs sampler. *Applied Statistics*, 43, 201–221.

Wakefield, J., Zhou, C., and Self, S. (2003) Modeling gene expression over time: curve clustering with informative prior distributions. In J.M. Bernardo, M. Bayarri, A.P. Dawid, J.O. Berger, D. Heckerman, A.F.M. Smith and M. West (eds), *Bayesian Statistics 7*. Oxford: Oxford University Press.

Walker, S., Damien, P., Laud, P., and Smith, A.F.M. (1999) Bayesian nonparametric inference for random distributions and realted functions. *Journal of the Royal Statistical Society, Series B*, 61, 485–527.

Wang, Y., Miao, Z.H., Pommier, Y., Kawasaki, E.S., and Player, A. (2007) Characterization of mismatch and high-signal intensity probes associated with Affymetrix genechips. *Bioinformatics*, 23(16), 2088–2095.

Watson, J., and Crick, F. (1953) Genetical implications of the structure of deoxyribonucleic acid. *Nature*, 171, 964–967.

Watson, J., Baker, T., Bell, S., Gann, A., Levine, M., and Losick, R. (2007) *Molecular Biology of the Gene*, 6th edition. Menlo Park, CA: Benjamin/Cummings.

Weaver, D.C., Workman, C.T., and Stormo, G. D. (1999) Modeling regulatory networks with weight matrices. *Proceedings of the Specific Symposium on Biocomputing*, pp. 340–359.

Wei, L. (1992) The accelerated failure time model: a useful alternative to the Cox regression model in survival analysis. *Statistics in Medicine*, 11, 1871–1879.

Wei, Z. and Li, H. (2007) Nonparametric pathway-based regression models for analysis of genomic data. *Biostatistics*, 8, 265–284.

Weigelt, B., Horlings, H.M., Kreike, B., Hayes, M.M., Hauptmann, M., Wessels, L.F., de Jong, D., van de Vijver, M.J., Van't Veer, L.J., and Peterse, J.L. (2008) Refinement of breast cancer classification by molecular characterization of histological special types. *J. Pathol.*, 216(6), 141–150.

Weir, I.S., and Pettitt, A.N. (2000) Binary probability maps using a hidden conditional autoregressive Gaussian process with an application to Finnish common toad data. *Applied Statistics*, 49, 473–484.

West, M. (2003) Bayesian factor regression models in the 'large p, small n' paradigm. In J.M. Bernardo, M. Bayarri, A.P. Dawid, J.O. Berger, D. Heckerman, A.F.M. Smith and M. West (eds), *Bayesian Statistics 7*, pp. 723–732. Oxford: Oxford University Press.

West, M., Muller, P., and Escobar, M. (1994) Hierarchical priors and mixture models with application in regression and density estimation. In A.F.M. Smith and P.R. Freeman (eds), *Aspects of Uncertainty: a tribute to D.V. Lindley*, pp. 363–386. Chichester: John Wiley & Sons, Ltd.

West, M., Blanchette, C., Dressman, H., Huang, E., Ishida, S., Spang, R., Zuzan, H., Olson, J.A. Jr, Marks, J.R., and Nevins, J.R. (2001) Predicting the clinical status of human breast cancer by using gene expression profiles. *Proc. Natl. Acad. Sci. USA*, 98(20), 11462–11467.

Westfall, P.H. and Young, S.S. (1993) *Resampling Based Multiple Testing*. New York: John Wiley & Sons, Inc.

Whitfield, M., Sherlock, G., Saldanha, A., Murray, J., Ball, C., Alexander, K., Matese, J., Perou, C., Hurt, M., Brown, P., and Botstein, D. (2002) Identification of genes periodically expressed in the human cell cycle and their expression in tumors. *Mol. Biol. Cell*, 13, 1977–2000.

Whittaker, J. (1990) *Graphical Models in Applied Multivariate Statistics*. Chichester: John Wiley & Sons, Ltd.

Whittemore, A.S. (2007) A Bayesian false discovery rate for multiple testing. *Journal of Applied Statistics*, 34, 1–9.

Wigner, E.P. (1955) Characteristic vectors of bordered matrices with infinite dimensions characteristic vectors of bordered matrices with infinite dimensions. *Ann. Math.* 62, 548–564.

Wilkinson, D.J. (2007) Bayesian methods in bioinformatics and computational systems biology. *Briefings in Bioinformatics*, 8, 109–116.

Wilkinson, D.J. and Yeung, S.K.H. (2002) Conditional simulation from highly structured Gaussian systems, with application to blocking-MCMC for the Bayesian analysis of very large linear models. *Statistics and Computing*, 12, 287–300.

Williams, C., and Barber, D. (1998) Bayesian classification with Gaussian priors. *IEEE Transactions on Pattern Analysis and Machine Intelligence*, 20, 1342–1351.

Wilson, D.L., Buckley, M.J., Helliwell, C.A., and Wilson, I.W. (2003) New normalization methods for cDNA microarray data. *Bioinformatics*, 19(11), 1325–1332.

Winter, S.S., Jiang, Z., Khawaja, H.M., Griffin, T., Devidas, M., Asselin, B.L., Larson, R.S., and Children's Oncology Group (2007) Identification of genomic classifiers that distinguish induction failure in T-lineage acute lymphoblastic leukemia: a report from the Children's Oncology Group. *Blood*, 110(5), 1429–1438.

Wu, Z., Irizarry, R.A., Gentleman, R., Murillo F.M., and Spencer, F. (2004) A Model Based Background Adjustment for Oligonucleotide Expression Arrays. *Johns Hopkins University, Dept. of Biostatistics Working Papers*. Working Paper 1. http://www.bepress.com/jhubiostat/paper1.

Xu, X., Olson, J., and Zhao, L. (2002) A regression-based method to identify differentially expressed genes in time course studies and its application to inducible Huntington's disease. *Hum. Mol. Genet.*, 11, 1977–1985.

Yang, Y.H., Buckley, M.J., and Speed, T.P. (2001) Analysis of cDNA microarray images. *Briefings in Bioinformatics*, 2(4), 341–349.

Yang, Y.H., Dudoit, S, Luu, P., Lin, D.M., Peng, V., Ngai, J., and Speed, T.P. (2002a) Normalization for cDNA microarray data: a robust composite method addressing single and multiple slide systematic variation. *Nucleic Acids Research*, 30(4), e15.

Yang, Y.H., Buckley, M.J., Dudoit, S., and Speed, T.P. (2002b). Comparison of methods for image analysis on cDNA microarray data. *Journal of Computational and Graphical Statistics*, 11(1), 108–136.

Yauk, C.L., Berndt, M.L., Williams, A., and Douglas, G.R. (2004) Comprehensive comparison of six microarray technologies. *Nucleic Acids Research*, 32, e124.

Yeang, C.H., Ramaswamy, S., Tamayo P., Mukherjee, S., Rifkin, R.M., Angelo, M., Reich, M., Lander, E., Mesirov, J., and Golub, T. (2001) Molecular classification of multiple tumor types. *Bioinformatics*, 17(Supplement 1), S316–322.

Yekutieli, D., and Benjamini, Y. (2001) The control of the false discovery rate in multiple testing under dependency. *Annals of Statistics*, 29(4), 1165–1188.

Yuan, M., and Lin, Y. (2006) Model selection and estimation in regression with grouped variables. *Journal of the Royal Statistical Society, Series B*, 68, 49–67.

Yuan, M., Kendziorski, C., Park, F., Porter, J.R., Hayes, K., and Bardfield, C.A. (2006). Hidden Markov models for microarray time course data in multiple biological conditions. *Journal of the American Statistical Association*, 101, 1323–1332.

Zellner, A. (1986) On assessing prior distributions and Bayesian regression analysis with g-prior distributions. In P.K. Goel and A. Zellner (eds), *Bayesian Inference and Decision Techniques: Essays in Honor of Bruno de Finetti*, pp. 233–243. New York: Elsevier.

Zhang, K., and Zhao, H. (2000) Assessing reliability of gene clusters from gene expression data. *Funct. Integr. Genomics*, 1(3), 156–173.

Zhang, L., Miles, M.F., and Aldape, K.D. (2003) A model of molecular interactions on short oligonucleotide microarrays. *Nature Biotechnology*, 21(7), 818–821.

Zhang, M., Yao, C., Guo, Z., Zou, J., Zhang, L., Xiao, H., Wang, D., Yang, D., Gong, X., Zhu, J., Li, Y., and Li, X. (2008) Apparently low reproducibility of true differential expression discoveries in microarray studies. *Bioinformatics*, 24(18), 2057–2063.

Zhao, P., and and Yu, B. (2006) On model selection consistency of lasso. *Journal of Machine Learning Research*, 7, 2541–2563.

Zhou, B.B., Peyton, M., He, B., Liu, C., Girard, L., Caudler, E., Lo, Y., Baribaud, F., Mikami, I., Reguart, N., Yang, G., Li, Y., Yao, W., Vaddi, K., Gazdar, A.F., Friedman, S.M., Jablons, D.M., Newton, R.C., Fridman, J.S., Minna, J.D., and Scherle, P.A. (2006) Targeting ADAM- mediated ligand cleavage to inhibit HER3 and EGFR pathways in non-small cell lung cancer. *Cancer Cell.*, 10(1), 39–50.

Zhou, X., Liu, S., Kim, E.S., Herbst, R.S., and Lee, J.J. (2008) Bayesian adaptive design for targeted therapy development in lung cancer – a step toward personalized medicine. *Clin. Trials*, 5(3), 181–193.

Zhu, G., Spellman, P.T., Volpe, T., Brown, P.D., Botstein, D., Davis, T.N., and Futcher, B. (2000) Two yeast forkhead genes regulate the cell cycle and pseudohyphal growth. *Nature*, 406, 90–94.

Index

DATE DUE